Emilio Carnevale

Sensors & Transducers

A practical guide

2023 edition

SUMMARY

1   INTRODUCTION

This book talks about sensors: inductive, capacitive, magnetic, optical, vision systems, encoders, and others. Sensors to detect strength, position, level, shape, or color. The operating principles and construction characteristics are explained. Practical examples are given, and typical applications are shown with a guide to choose. It is suggested how to draw sensors in electrical patterns and how to electrically connect them.

What's in this book that I couldn't find on the Internet? Nothing. But there isn't a website on the Internet that treats all these devices in a single document that's easy and quick to consult and full of images.

Aside from the signal and interface displays, the various controls and buttons that are required for the different functions, what allows a machine to interact with the real world are the sensors. A machine "sees" using a camera, "hears" the presence of an object with a sensor, detects an abnormal condition as too high a temperature through a thermocouple. In addition, sensors allow the machine itself to know the position of its organs in motion, such as encoders in a robot arm or simple magnetic sensors that check whether a pneumatic actuator is in the forward or reverse position.

I worked for several years in the field of industrial automation, first as a technician and maintenance engineer and then as a designer. It was a long and formative experience and at the end I found myself with an incredible collection of catalogs, technical sheets and personal notes concerning the many machines I have treated.

I decided that the sensor literature was very interesting, and I didn't want to throw it away. This was the idea behind the book which originally wanted to be a collection, in a single text and in a homogeneous form, of all the knowledge gained in the use and choice of the various types of sensors.

The first edition of this book, entitled "Sensors and transducers for industry and automation - a practical guide" was published in 2008. To see it now, that little book makes you smile. In 2017, after nine years, I decided to tackle its revision. The second edition has been revised, corrected, and extended, also in the light of technological updates, regulations and experiences in other work and personal fields.

This edition has been further revised and corrected, the chapters have been reorganized to make the consultation more practical and less dispersed. This edition explicitly described and contextualized some sensors that were only hinted at or used in other contexts, such as accelerometers, sensors for digital cameras, and those used in modern smartphones.

This manual is not intended to replace a textbook, it continues to be a practical guide to understanding the operation and use of the various types of sensors that can be used by working in industry, in the field of industrial automation or in a laboratory.

The practical and schematic approach seeks to meet the real needs of designers and users of sensors and transducers. We tried to provide comprehensive information to guide you in choosing the right sensor to use it correctly, omitting unnecessary formulas and data wherever possible.

For those who need more in-depth guidance or need calculations for the sizing of a project, it is recommended to read specific texts and consult the technical and application sheets and catalogs of the various manufacturers.

The sensors described are generic sensors, no manufacturer is mentioned. Most of the features of the same type of sensor are similar, regardless of manufacturer. In some cases, there may exist special sensors or very different from the "standard" characteristics, sometimes made by a single manufacturer.

As this sector is constantly and constantly evolving, it is advisable to check for different or better sensors than those described in this guide before undertaking a specific project.

In my work experience and as a hobbyist I have had the opportunity to use, or at least test, many of the sensors described here. I hope this work will be useful to technicians and designers.

Any suggestions, corrections or reports will be welcome, please send your comments to: emilio_carnevale@hotmail.com.

1.1   Symbology

The following are the various symbols that appear in the text with their explanatory notes used to complete the text.

Regardless of the notes that can be found in this book, we recommend that you strictly follow the instructions and warnings provided by the manufacturer of the sensor used.

DANGER: danger to persons (e.g.: LASER sensors or sensors that can operate at high voltages).

ATTENTION: Sensor damage capability (connections, mechanical stress, over range, etc.).

INFORMATION: notes for proper use. If not observed, no damage occurs, but the sensor may not work in optimal conditions.

Notes and Curiosity.

1.2   Abbreviations

°C   degree Celsius or centigrade, unit of temperature

°F   degree Fahrenheit, unit of temperature

A   Ampere, unit of electric current

AC   Alternate Current, AC

CAD   Computer Aided Design, PC-assisted drawing

CCD   Charge Coupled Device

CEI   Italian Electrotechnical Committee

cm   centimeter

DC   Direct current, direct current

FDA   Food and Drug Administration

IEC   International Electrotechnical Commission

kHz	kilohertz (thousands of oscillations per second)
LED	Light Emitter Diode
mV	millivolt (millisecond of Volt)
nm	nanometer (billionth of meter)
PC	Personal Computer
PID	proportional/supplementary/derivative
PLC	Programmable Logic Circuit
Range	interval between two values or quantities
RC	resistance-capacitance
RGB	Red-Green-Blue, Red-Green-Blue
RTD	Temperature Detector Resistance, Thermoresistance
UV	ultraviolet
V	Volt, unit of electric potential difference

1.3   Technical terms

The following is a reasoned list of terms which can be found in the manufacturers' catalogs and technical sheets. Some terms are only valid for a certain type of sensor, e.g., the detection distance makes no sense in a pressure transducer.

- **Current absorption:** also called **self-consumption**, is the consumption of the sensor fueled and in a state of rest (off-line output).
- **Voltage drop:** the difference between the supply voltage and the output voltage with the sensor in conduction (active).
- **Measurement range**: the space in which the object to be measured must be placed in such a way that it has a stable reading in the declared parameters. Some sensors have a fixed detection distance.
- **Maximum output current**: maximum current that can be drawn from the sensor.
- **Maximum current of cue**: the maximum current the sensor can withstand for a limited time.
- **Minimum output current**: minimum current value for reliable operation.
- **Residual current**: the current through the sensor with the output off.
- **Thermal drift**: maximum variation of the intervention distance (Sn) within permitted temperature limits, expressed as a percentage of +/- 10% Sn.

- **Operating distance**: Typically indicated in proximity sensors, it is sometimes denoted Sn. For inductive purposes, it is a square metal plate (Fe 360) with a side equal to or greater than the diameter of the active surface of the sensor and a thickness of 1 [mm]. It is measured at 20°C.
- **Working frequency**: maximum number of switches per second.
- **Degree of protection**: the protection achieved by the electrical part container. In inductive and capacitive sensors, the first digit is often 6 (total protection against dust), the second digit can range from 5 (protection against spraying of water) to 7 (immersion for a fixed period).
- **Hysteresis**: for an ON-OFF sensor, the difference between the excitation and de-excitation values changing the target distance. As a general term, hysteresis means a lag between input and output in a system upon a change in direction. It can be expressed as an absolute value or as a percentage of Sn.
- **Temperature limits**: temperature range within which the declared operating conditions are guaranteed.
- **Residual undulation**: is the percentage of righted alternating voltage, measured peak-to-peak, overlaid with the continuous supply voltage of the sensor. Generally, 15% residual undulation is allowed for sensors in DC.
- **Protection against voltage peaks**: protection against damage from extra tension due to inductive peaks.
- **Signal-to-noise ratio** (SNR): This feature highlights the useful signal power over noise power in any information acquisition, processing, or transmission system, usually in decibels.
- **Insulation resistance**: resistance expressed in ohms, between the sensor circuits and the metal container, measured by applying an alternating voltage of 500 [V].
- **Repeatability**: is the variation that can undergo the intervention distance between two consecutive actuations of the same sensor on equal working conditions.
- **Resolution**: The smallest variation of measurement that can be appreciated by the sensor.
- **Response delay**: is the time between detection and switching of output.
- **Temperature stability**: indicates the percentage of error in the unit of temperature measure (°C). In practice, this results in a deviation of the measurement at different temperatures. Temperature stability is

important for accurate measurements, especially in industrial applications where large variations can occur.

- **Ambient temperature**: the temperature at which the technical specifications are guaranteed.
- **Response time**: the time taken for the output signal to move from 10% to 90%.
- **Rated voltage**: A voltage range that ensures proper operation. Do not exceed the specified maximum voltage.
- **Sampling speed**: is the rate at which the analog signal is sampled during analog/digital conversion. That is, it indicates how long a sample of the size to be measured is detected.

The terms "sensor" and "transducer" are often synonymous and are commonly used indiscriminately.

It's not wrong to define a sensor as a transducer, but basically the transducer is also a sensor, in the sense that it senses something and communicates it through an electrical signal. In common usage, the terms sensor and transducer are often equivalent, and it is customary to call a sensor a general transducer. The opposite is not true.

However, the different names suggest that these two categories of components are working differently.

Generalizing enormously, we can say that a SENSOR is a device that detects the presence or characteristics of a target and gives an output "on or off" signal. A transducer converts a value of a physical size into a proportional electrical signal.

Sensors are devices that can detect a magnitude of various kinds such as force, temperature, speed, distance, proximity of objects and can interact with an electrical system signaling a change in measured magnitude such as the presence of objects to be picked up or the achievement of the working position of an actuator.

Transducers provide an electrical signal proportional to the size value affected, useful if you want to implement continuous control of a process with the possibility to perform corrective actions of regulation, for example the regulation of the water temperature in a boiler. They can also be used simply to measure a size as in a common digital clock with built-in barometer and thermometer.

The use of sensors allows a plant with appropriate logic, such as computers, PLC or dedicated electronics, to control any kind of industrial process and to have under control all the quantities necessary for the automatic management of the process itself.

There are many different types of sensors designed for different uses. It is therefore important to be able to choose the type best suited to your needs and use it correctly.

Sensors with complex output signals and transducers with a very different output signal from a simple analog signal are available.

Other devices that cannot fit into simple definitions or subdivisions are e.g. barcode readers or vision systems. They are dealt with in a specific chapter.

In the next few chapters, we look at different types of sensors and transducers, trying to understand their principle of operation and their use. Some operating principles are explained in the specific chapter on sensors, the same principles are obviously also valid for transducers.

2.1   Sensors with on-off output

These sensors, also called **logic output** sensors or **all-or-nothing** sensors, have the characteristic of having a switching of the output on reaching a fixed threshold value.

For example, as the pressure increases beyond the calibration threshold, a typical pressure switches its output. Or the same happens to a capacitive sensor as an object approach.

In some sensors this threshold is fixed, a typical case is a standard inductive sensor, in others it may be adjustable, as is generally the case with a pressure switch.

Output switching generally results in the voltage being passed on the output conductor, generally a BLACK wire in three- or more-wire sensors, from a value of 0V to the supply voltage value, typically 24V.

We can distinguish between sensors with **NPN** or **PNP** output. In the sensors with NPN output we have the output that leads to the value of 0V, referring to the positive supply voltage. These sensors are less common than PNPs but are still present in most manufacturers' catalogs. Some sensors exist only in the NPN version.

In sensors with PNP output, however, we have the output that goes to the value of the supply voltage (24V), referring to the mass (0V) or negative supply voltage.

An ON-OFF sensor may also have an output composed of a **mechanical contact**, such as a reed sensor, a thermostat, a fluxsostat. Generally, these sensors do not require a supply voltage and the output contact can be used directly in a circuit, such as a relay contact. It is possible to use the sensor in both PNP and NPN logic plants.

Generally, sensors also have an LED indicating the readout condition, i.e., switching the output.

It is possible that the output circuit is not the same as that of the signal LED. The signal LED may therefore function correctly, but the sensor output may not function and vice versa. In doubt check the output with a tester and do not trust the LED.

Do not exceed the maximum permissible load on the output circuit of a sensor (typically several tens of mA). This value can be found on technical sheets and catalogs of manufacturers.

A common phenomenon that can be detected using sensors is **hysteresis**. This essentially means that the excitation and de-excitation of a sensor does not both occur under the same detection conditions.

Let's take an example. We use an inductive sensor to detect a piece of metal, and we bring the metal closer to the sensor until it detects it and switches its output. Suppose the switching took place at a sensor-metal distance of 3 [mm]. We continue now to approach the metal to the sensor, of course the output of the sensor does not change, then we begin to move it away, until the output returns to the resting state. We can detect that the deexcitation of the sensor occurred at a distance slightly greater than 3 [mm].

Hysteresis is a normal phenomenon, common to most sensors, which in practical use usually does not cause problems but should be considered when precise detection is required.

2.2 Analog output transducers

These sensors, also called transducers, have an output signal proportional to the measured input size that can be, e.g., pressure, temperature, or distance. They are also referred to as measurement sensors. A second ON-OFF type output can also be provided in combination with the analog output.

Typically, the output of an analog sensor is a **voltage value** that can vary from 0 to 10V, or a current value of 4 to 20mA. There are however many different output circuits with different voltages and currents (0-5V, 0-100mV, 0-20mA). The output voltage generally refers to the supply voltage, but the sensor can also have a separate and isolated output circuit.

This type of sensor can be used for a variety of applications with measurement requirements. Unlike ON-OFF sensors which can only detect whether an event has occurred, analog output sensors provide a value that varies with the process being measured, allowing control and adjustment of the process.

A typical example might be an oven controlled by a temperature sensor, thermoresistance type PT100 or something, and an intelligent regulating device such as an advanced thermoregulator or dedicated electronics, which could implement a PID algorithm to keep the temperature extremely constant over time.

Ideal and real transducers

An **ideal** transducer has a perfectly **linear** output characteristic, usually a line that intersects the Cartesian plane at the origin and proceeds unaltered to the maximum allowed value at the transducer's input which usually also corresponds to the maximum amplitude of the output. An output signal also equal to 0 should correspond to an input 0 magnitude.

In practice, this condition is impossible, and a common transducer has several **typical errors** and **deviations from the ideal characteristic** to be considered.

One possible error that characterizes transducers is that of non-linearity or deviation from the ideal line, which can be systematic or random.

Systematic **non-linearity** is caused by the principles of operation and the construction technology of a given transducer, usually repeated in every specimen of the same type, and can partly be compensated with suitable dedicated algorithms or circuits.

Random nonlinearity is caused by inaccuracies in construction and by characteristic drifts over time. This type of nonlinearity is present to a greater or lesser extent in any type of transducer and cannot be compensated for.

Another typical error is **offset**. In practice, at an input magnitude of 0, the output of the transducer is not exactly 0. The offset error is expressed as the output value with zero input. This error is typically easily compensated by software or hardware.

Gain error is another variation from the ideal characteristic of a transducer. The gain error in practice results in a different inclination of the real line, compared to the ideal line. The line represents the output signal of the transducer.

The gain error is usually expressed as a percentage. This error is also easily compensated.

There are also drifts and variations of the signal due to room **temperature**. Transducers manufacturers provide typical values of their components referring to a specific temperature (20-25 °C), the components usually exhibit behavior variations as temperature changes.

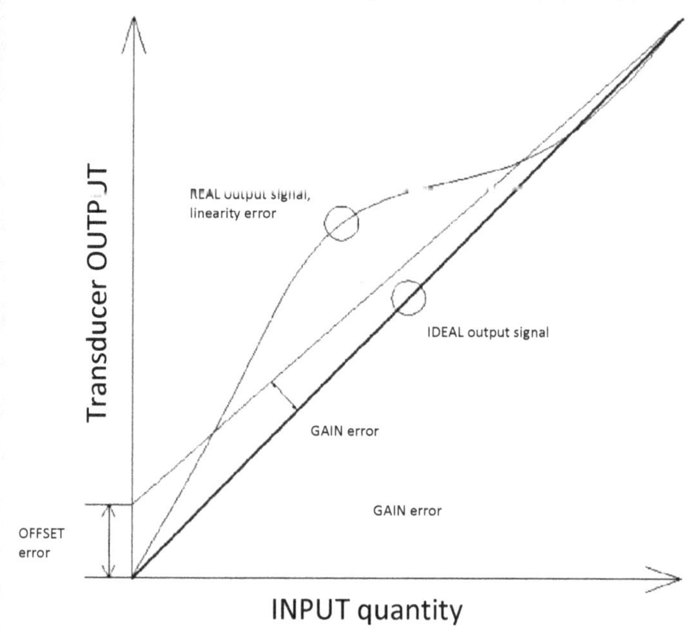

Other errors include **quantization** errors, which are basically the maximum value that can be applied to the transducer without its output indicating variations, and hysteresis which is the difference of the output value, for the input signal equal, between the measure obtained by increasing the input value and the measure obtained by decreasing it.

Quantization is sometimes referred to in technical catalogs as resolution.

For good quality components to express variations of typical parameters, especially variations due to temperature, with a value in percentage is impractical, as they would result in very small numbers. In these cases, the indication in **ppm** (parts per million) is used, in practice it is the percentage value multiplied by 10,000.

Example: We assume that we have a transducer with an offset temperature coefficient of 3 µV/°C and an offset of 200 µV. An ambient temperature change of +40 °C will result in an offset increase of 120 µV (3 µV/°C x 40 °C), so its overall offset will now be 200+120=320 µV.

The following is a list of some significant parameters that are usually found in transducers catalogs and technical sheets:

- **Range**: also called range, is the range of measurable values, for a given quantity. In this range the transducer works as specified.
- **Transfer function**: is the relationship between the measured magnitude and the output signal. Usually in catalogs it can be expressed as graphs or tables, but formulas can also be used.
- **Response time**: indicates the reactivity or velocity of the transducer in responding to changes in the magnitude to be measured. The response time is indicated as the time taken for the component to vary the output against an input change of 90 or 95% of the measurable range.
- **Sensitivity**: is the ratio of the change in the measurement, output, to the change in the input quantity.
- **Resolution**: represents the smallest measurable variation in input that can be detected on the output signal.
- **Linearity**: as we've seen, all transducers are not ideal, and their output is not a line, but a curve. The linearity error is the maximum deviation of the output signal from the ideal line. It is usually expressed in %.
- **Stability**: indicates the sensor's ability to maintain its characteristics over time.
- **Hysteresis**: represents the difference of the output value by measuring the same magnitude. It can be measured by detecting the output value, first increasing the input magnitude, and then decreasing it.
- **Output offset**: the value of the signal, which should be zero, present on the output circuit by measuring a magnitude of 0% of the measurable range, as shown in the previous image.

Other parameters that classify a transducer based on its behavior over time, by performing several measurements of the same size:

- **Accuracy:** compares the measured values of the test component with those of its ideal characteristic, usually expressed in %.
- **Precision** (repeatability): refers to the transducer's ability to output values very close to each other in a series of close tests performed with the same input value.

There may be very precise but inaccurate transducers, in which a set of measurements yields results that are similar but not equal to the ideal value, or economic transducers that can be both inaccurate and inaccurate. Generally, a precise and accurate component is also very expensive, and it is worthwhile to assess whether its use is necessary.

The image below can make understanding these behaviors more immediate. Each dot represents a measurement detected by the transducer output signal, measuring the same size. The center of the 'target' represents the ideal result:

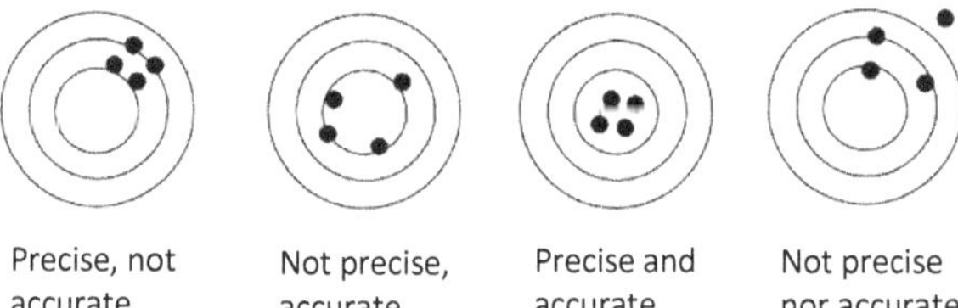

A transducer can be very precise, meaning that its measurements are all similar, but inaccurate if the result is very different from the ideal center. An inaccurate transducer, on the other hand, provides very different measurements even when measuring the same size.

Transducers can also be classified according to precision, and are divided into classes for this purpose:

Class	Full-scale error	Usage
5 – 2.5	± 5% - ± 2.5%	Normal use
1.5	± 1.5%	Precision
≤ 1	≤ ± 1%	Precision, testing and laboratory
≤ 0.1	≤ ± 0.1%	Reference transducers

3 SENSORS

3.1 Inductive sensors

The inductive sensor is based on the interaction of **metal objects** with the electromagnetic field of the sensor. Parasitic currents are induced in the metallic body, so the electromagnetic field is subtracted, and the sensor signals the presence of the object through its output.

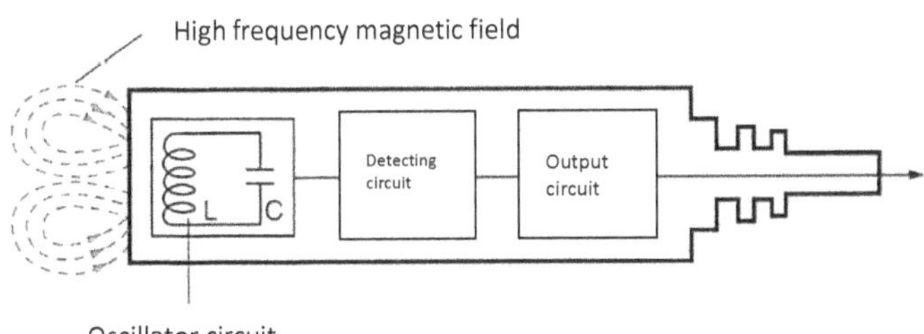

The structure of a typical inductive sensor is composed of a coil, which is part of an **oscillator** circuit, located near the reading area, a **trigger** circuit and an **output** circuit that generates the pick-up electrical signal for use.

In practice, the oscillator generates a high frequency magnetic field, when a metal body approaches the electromagnetic field and a current start flowing on its surface, this current subtracts energy from the field and causes the amplitude of the oscillations to decrease as the metal approaches the sensor, until the oscillations stop. The trigger circuit then detects oscillations and activates the output circuit, which is usually composed of a transistor in DC sensors.

Inductive sensors are also called **proximity** sensors. The structure of a typical inductive sensor is a cylindrical threaded body, usually made of metal. The power cord and the output signal shall either be attached to the sensor body itself or connected via a connector.

Type of construction

The typical form of an inductive sensor is cylindrical, smooth, or threaded, typically M4 to M30, and generally metal (chromed brass, steel). The reading area, i.e., the end of the cylinder, is usually of plastic material.

Fixing is done for threaded sensors with the nuts supplied using a hole or a dedicated bracket usually supplied as an optional. The latter solution allows the precise orientation and positioning of the sensor itself, as the brackets have specific holes and soles.

There are both shielded and non-shielded models. A sensor shielded in practice is all-metal throughout its length, with the plastic reading area flush of the container.

Shielded sensor, left, and unshielded sensor with related schematic magnetic fields:

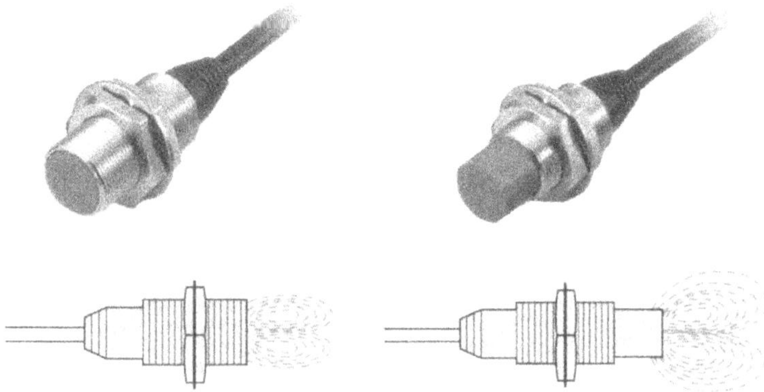

The unshielded sensor has a sensitive end of plastic material protruding from the metal body.

A shielded sensor can be fully housed in a metal hole without affecting reading. Always use shielded sensors if there are metals in place in the reading area or for a wire mount.

An example of a fastening is shown. The screened sensor, on the left, can be mounted to wire of a metal flange, note the plastic end, without the thread, of the unshielded sensor:

There are inductive sensors in other forms as well, such as parallelepipeds or flattened sensors, to be inserted into tight spaces. Another possible form is the toroidal form, used for example for the detection of passage of metal pieces into a plastic tube or falling. These non-cylindrical sensors are generally made of plastic.

Typical technical features
- Power supply voltage 10 to 30 Vdc
- PNP or NPN output
- Sn, variable reading distance based on sensor size from <1 [mm] to >10 [mm]

Typical applications

Detection of presence, proximity, of metal objects. Position detection and end travel of moving organs. Check for parts, such as passing screws for fall into a plastic tube, with a ring inductive sensor.

The following table shows the nominal switching distance *Sn* for reading metal objects. The reference is the iron with which the inductive sensor achieves the best performance in terms of detection distance.

Sn value	Material
1	Iron FE 360
1	Steel / Cast iron
0,6 – 0,9	Stainless steel
0,7	Nichel
0,3 – 0,4	Copper / Aluminum
0,4 - ,05	Brass / Bronze

The nominal reading distance **Sn** in catalogs refers to an **iron** target. If a metal other than iron is to be detected, consider that the reading distance decreases, as shown in the table.

3.1.1 NAMUR sensors

NAMUR is the initials of the German Committee for Standards and Control. NAMURs are sensors, but it would be better to call them transducers, unamplified **inductive**, two-wire, polarized DC current.

They change their internal resistance depending on the **distance** of the object to be detected, they contain only the oscillator.

They require only a few construction elements and therefore offer the best operational safety. Due to the low ohmic closure resistance, the sensor is insensitive to the inductive or capacitive dispersions present on the connection line with the control amplifier. Typical values are a supply voltage of about 8 [V], a resistance of 1000 [Ω] and currents in the order of 1 or 2 [mA].

They must be connected to an electronic amplifier circuit, such as the module shown in the figure, which converts the change in current signal to an ON-OFF type output. The set is therefore in all effects a sensor that switches the output state depending on the distance of the target detected.

Operating on the inductive principle, only metallic objects can be detected.

Often these sensors are blue in color for the read head or some other particulars such as the output cable.

3.2   Capacitive sensors

The capacitive sensor bases its operation on the capacity variation induced by the presence of a body near the reading area of the sensor. This change in capacity is therefore also caused by **non-metallic**, liquid, or other bodies.

The structure of a typical capacitive sensor consists of a transistor oscillator, the RC (resistor-capacity) **oscillating circuit** near the reading area, a **trigger** circuit, and an **output** circuit that generates the pick-up electrical signal for use.

The capacity consists of two plates located near the reading area of the sensor that usually use air as a dielectric. This variation induces oscillations in the oscillator circuit, which in turn activates the trigger and thus the output circuit, generally a transistor in DC sensors.

In the figure is shown the structure of a typical plastic housing capacitive sensor with threaded body, with the power cord and the output signal circuit directly housed in the sensor body:

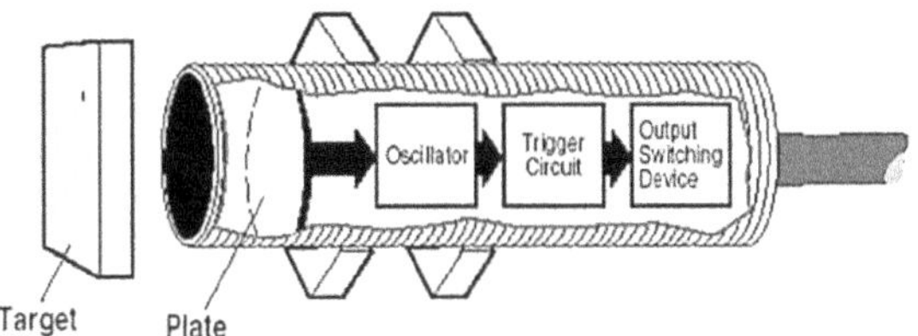

In the inductive sensor the oscillations are stopped by the approach of a metal, while in the capacitive sensor the oscillations are activated by the approach of an object. This makes capacitive sensors a little slower in response than the inductive ones.

Type of construction

The typical shape of a capacitive sensor is with a cylindrical threaded body of plastic material. The sizes can vary, generally the smaller sensors have M8 thread, the larger ones can be M30 or more.

Fixing is by using the supplied dice through a hole or bracket. The latter solution usually allows the precise orientation and positioning of the sensor itself, as the brackets have specific holes and soles.

Some capacitive sensors may have an amplifier separate from the read head unit.

Typical technical features

- Power supply 10 to 30 Vdc
- PNP or NPN output
- Variable read distance based on sensor size from <1 mm to >10 mm
- Sensitivity can be adjusted on some models

Typical applications

Non-metallic or non-magnetic object detection, liquid, and dust detection.

The nominal reading distance Sn found in catalogs refers to a METAL target, typically iron. If you want to detect a different material, you need to consider that the reading distance decreases. See the table below.

Nominal switching distance comparison table for non-metallic materials reading:

SN value	Material
1	Metal / water
0,5	Plastic / glass
0,2 – 0,7	Wood
0,1	Oil

3.3   Magnetic sensors

Some materials such as permanent magnets have the property of inducing a field in the space around them such that any nearby metal body will suffer a force. This field is called the magnetic field. A magnetic field can also be generated by electric charge motion; it does not change its nature whether it is generated by an electric current or generated by a permanent magnet.

In magnets one can always find two distinct and opposite poles, north N and south S as with two electric charges, similar poles repel, and different poles attract.

The magnetic sensor is composed of magnetic metal foils enclosed in a small glass tube filled with inert gas. The two foils relate to two conductors for the electrical connection. This structure is called **reed** contact.

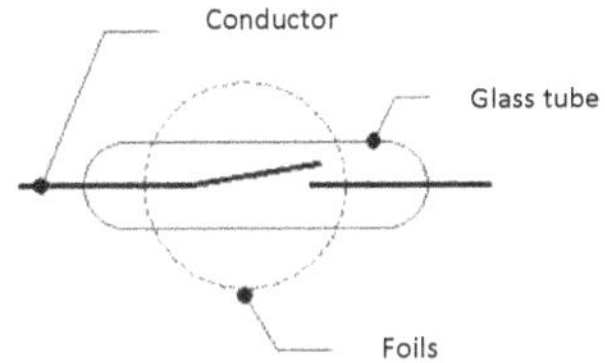

Metal sheets detect the proximity of magnetic fields.

A solid magnet, for example, on the movable stem of a pneumatic actuator, by the principle of magnetic induction generates, on the two foils of the reed contact, a polarity of opposite sign. When the external magnetic field generates a sufficient attraction force to overcome the elastic force of the foils, they flex and touching each other close the electrical contact.

Devices made with this operating principle shall be used for the detection of position, distance, speed, and acceleration. See also the chapter on magnetic-mode position transducers.

Examples of magnetic sensors for the detection of the stem position of pneumatic actuators, shaped to be fixed on the actuators themselves are shown:

The principle of magnetic operation is also based on the Hall effect sensors, explained in detail in section 4.10.3.

A Hall effect sensor varies its output voltage in response to a magnetic field. Hall effect devices are used as: proximity sensors, positioning, speed, and current detection. Unlike a mechanical switch, it is a long-lasting solution as there are no mechanical wear problems.

Type of construction

Hall effect sensors can have many shapes: cylinders, parallelepipeds or, in the case of sensors with pneumatic actuators, be shaped to allow mounting on the actuators themselves, in the designated locations.

Reed sensors can have the same construction shapes as Hall sensors.

Typically, these sensors have integrated electrical cable but there are also connector versions.

Magnetic sensors for safety applications exist, either cylindrical or parallelepiped with a dedicated magnet, usually of the same shape and color. The magnet can be coded, allowing operation only with its own sensor.

Typical technical features

- 5 to 30 Vdc power supply, with 2 or 3 wires connection
- NA or NC contact output

Typical applications

Positioning detection of pneumatic actuators with a magnet integral with the moving stem.

Safety applications.

The reed sensor metal foils may be coated with noble materials that give the possibility to drive directly other circuits or inductive loads. Compared to traditional mechanical limit switches, magnetic sensors are dust protected because the contacts are underwater, have a significantly longer duration and do not require maintenance.

3.4   Optical sensors

The optical sensor is also called **photoelectric, optoelectronic,** or commonly **photocell**.

The range of optical sensors is certainly the largest, comprising different technologies such as **visible light** or **LASER**, many different types of construction and special uses, such as color detection of an object.

All these sensors are based on light, which is electromagnetic radiation.

A typical optical sensor consists of:
- a photoemitter controlled by an electronic circuit that modulates light pulses
- an optical system for directing the beam
- a photo-receiver that converts the light received into electricity
- a demodulator circuit that amplifies the part of the signal that is generated by receiving the modulated light only
- a comparator comparing the signal obtained with a threshold, usually user settable
- an output circuit, usually a transistor

Follow the block pattern of a typical optical sensor with emitter and receiver in the same device:

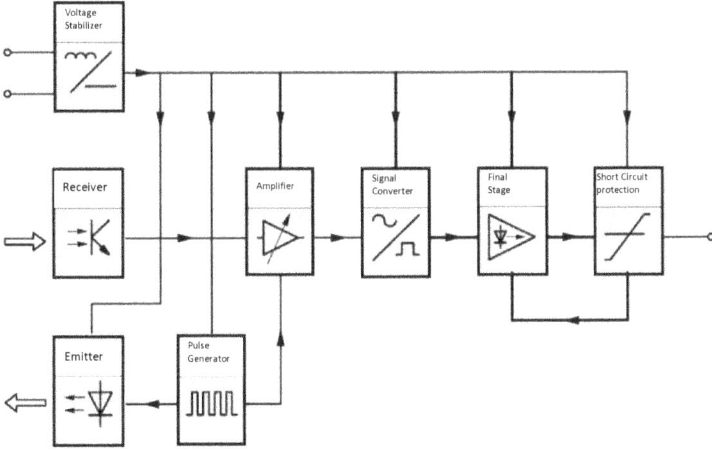

When light strikes an object, there are certain phenomena, such as **reflection**, **absorption**, or **transmission** to different degrees, depending on the characteristics of the object such as the type of surface, gloss, roughness, color, thickness in the case of transparent objects, distance. These features can be detected by an optical sensor.

The main features common to optical sensors are:
- Contactless detection
- Object detection of any material
- Long distance detection
- High speed response
- Possibility of color discrimination
- High precision detection of even small objects

Generally, an LED is used as the photoemitter. With modern technology, virtually all colors of visible light can be achieved with an LED, as can infrared and ultraviolet light. Red light or infrared LASER diodes are also used.

The emission of the light ray can be either continuous or pulse based. Continuous emission is characterized by vulnerability to ambient light and rapid response, while pulse emission can provide a long detection distance with minimal interference from ambient light.

Virtually all the all features described below are also available with fiber optic sensors.

here are optical sensors with LASER light emission. Be especially careful with your eyes. While conventional LASER sensors are not hazardous, they should be cautioned and not look at the beam directly. See the LASER paragraph in the APPENDIX chapter.

Optics overview

Visible light has a wavelength between approximately 380 [nm], corresponding to the violet color and 750 [nm] corresponding to the red color.

At wavelengths below 380 [nm], i.e., at higher frequencies, ultraviolet rays begin, while at the opposite, below red-light frequencies, infrared, or heat start, both not visible to the human eye.

Light is perceived as white if it is composed of all the components of the visible spectrum but appears instead colored when a specific wavelength range predominates over the others.

The light makes different interactions with the materials it strikes. These vary according to the wavelength of the emitted radiation, the interaction with the atomic structure of the material and its surface and the angle of incidence between the radiation and the material.

These interactions can be categorized into transmission, absorption, reflection, dispersion, and refraction.

Transmission is the permeability, transparency, to radiation of a given material. The vacuum, for example, is permeable to radiation while all other substances absorb to varying degrees. It follows that only a percentage of the energy emitted by a transmitter will reach the receiver.

Absorption is achieved by the total or partial transformation of the radiation received into another form of energy such as heat, or into a different type of radiation such as the different wavelength radiation used by luminescence sensors.

Reflection is the phenomenon that occurs when the radiation affecting a material is reflected, or re-emitted, and returned to the environment, in whole or in part.

Reflection can occur in all directions, even towards the same emitting source. The reflectivity or reflectance or reflection factor R of a surface, at a given wavelength, represents the percentage of energy that is reflected, compared to that absorbed.

Reflection can occur at well-defined angles, and mirror reflection occurs when light is reflected at an angle equal to the angle of incidence, with respect to the surface of the material.

If light hits a rough or irregular surface, it is reflected in several directions simultaneously, in which case it is possible to speak of **diffusion**.

Material	R Reflection Factor
Silver	0,88-0,93
Aluminum	0,55-0,7
Steel	0,55
White paint	0,6-0,7

Materials are characterized using three factors: T transmission, A absorption, R reflection. The sum T+A+R is always 1, which is 100% of the energy received.

Dispersion occurs when light changes in the direction of its path within the same material, due to a different refraction of the various colors in the spectrum. By passing a beam of light through a triangular-shaped glass prism, the light at the prism's output is decomposed into the basic colors of the spectrum.

This dispersion phenomenon is explained through refraction. The speed of light in a medium such as glass differs slightly depending on its wavelength. Since at different wavelengths light has different colors, and the refractive index depends on the speed of light in the middle, at the output of the prism the light beam will have slightly different angles of refraction depending on the color: the smaller the wavelength, the greater the angle of refraction.

Refraction is the deviation that a bright ray, inclined to the surface of the impacted material, undergoes in passing from one transparent medium to another. For example, we can mention the air-glass passage, the classic prism.

Beyond a certain angle of incidence, however, light can no longer penetrate the second material, in this case it is called total reflection. For the air-glass case this occurs at an angle of 42°.

Electromagnetic radiation

The following table lists some of the most common types of light radiation and the corresponding wavelength.

Wavelength [nm]	Light type
100 - 280	Ultraviolet C
280 - 315	Ultraviolet B
315 - 380	Ultraviolet A
380 - 455	Purple
455 - 495	Blue
495 - 558	Green
558 - 597	Yellow
597 - 622	Orange
622 - 750	Red
750 - 1400	Infrared A

The color of light in optical sensors

Optical sensors, usually those using **fiber optic** with separate amplifier, can be available with light emission different in color from the standard red light.

The availability of different colors is motivated by the fact that light is reflected and absorbed differently depending on the color of the object. Furthermore, some specific shades of a color may be difficult to detect with certainty, it is therefore appropriate to evaluate the use of an emitter emitting light with a different wavelength, i.e., a different color.

Some manufacturers provide tables in their catalogs to help choose the color of light to use. Typically, red is available as standard and colors as WHITE or GREEN as an alternative.

The following graphs are an example of the amount of light energy returning to the sensor depending on the color of the target (or background!), shown below, and the color of the light emitted, in percentages.

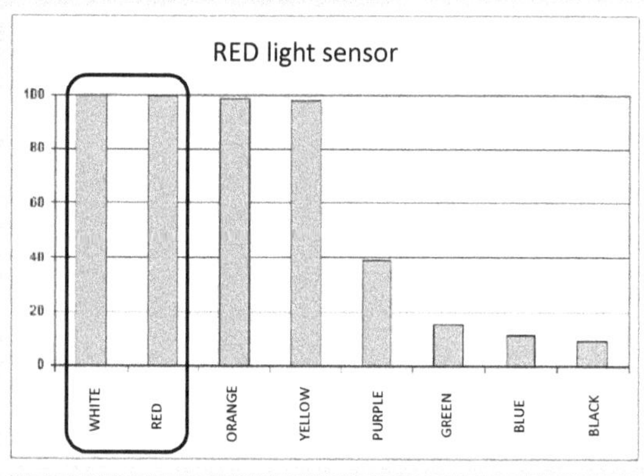

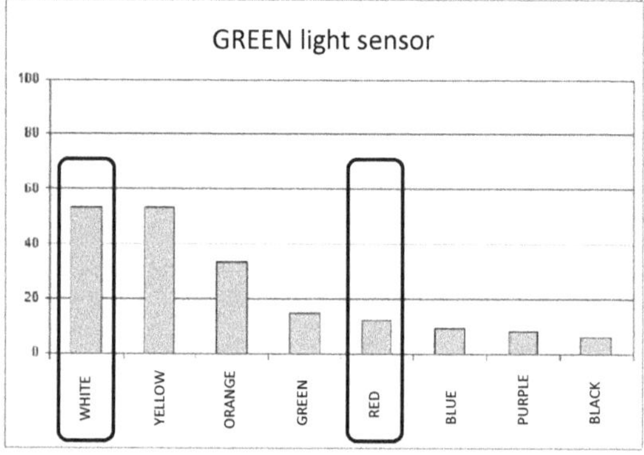

The graphs clearly show that if I want to safely detect a red object on a white background or vice versa, I must use a green light sensor.

The green light sensor has very different values in reception, and it discriminates very well against RED from WHITE, while for the red-light sensor there is practically no difference, so I would not be able to calibrate it.

Assuming you have an ORANGE color object on a GREEN background or vice versa, it is better to use a red light because the green light sensor would have difficulty distinguishing the two colors.

The same is true for common pairs of colors such as RED and BLUE, for which the red light sensor is always recommended.

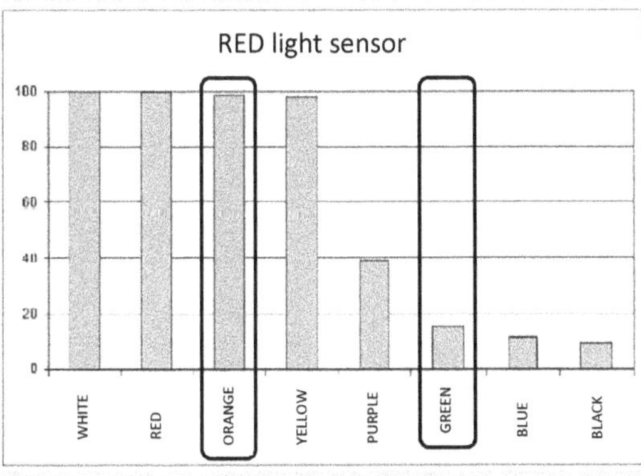

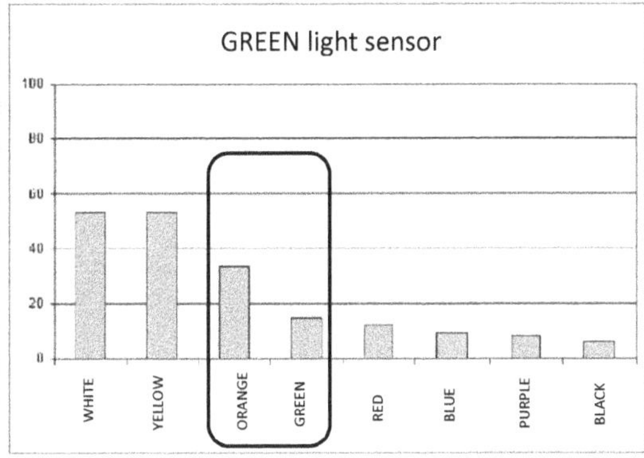

3.4.1 Optical fibers

All the major optical sensor functions are also fully implemented using fiber optics technology. Optical fibers can be assimilated to electrical cables that carry light energy instead of electricity. Their characteristic is **to carry a light pulse** at a great distance without a significant loss of signal intensity, taking advantage of the principle of total reflection.

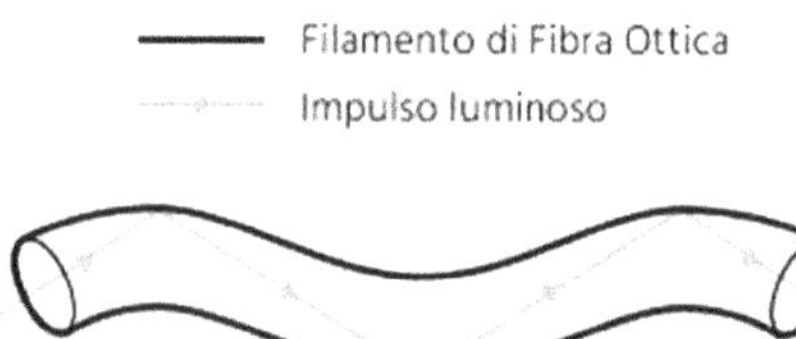

The world of fiber optics is vast and diverse. There are fibers for multiple applications and made from different materials.

Optical fibers are not only used to make sensors but are also used in other fields such as telephony and hi-fi video and audio. In addition, due to the lack of electricity transport, they can be used in dangerous environments and explosive atmospheres.

Essentially an optical fiber is a tube, cylindrical central **core**, made of glass or plastic material, surrounded by a protective **mantle**. The difference between the refractive index of the core and that of the coating forces light to travel through the core without being dispersed outside. The entire fiber is covered with a protective sheath made of plastic or even metal.

The inner core that allows the passage of light can be made with a single fiber or using many microfibers, thus making objects much more resistant to flexions and bends.

The materials with which the fibers are coated can be heat resistant, such as glass fiber coated with steel, or resistant to chemical agents with Teflon®-type special plastic sheath, or extremely flexible to allow use in application with moving mechanical parts. They can also be shielded cable stainless steel braid or spiral.

The most common types of fiber are:
- Plastic. Acrylic resin fiber enclosed in a polyethylene sheath. The fiber is light and flexible, which is the most common type
- Glass fiber core protected by a steel casing. Usable in high temperature zones up to approximately 350°C

The fiber end can have many shapes, usually consisting of a threaded metal termination, M4 and M6 the most common filets, in axis with the fiber itself or even angled 90°, to save space and to avoid abrupt folds of the fiber itself.

In particular models, special metal terminals allow you to shape the termination of the fiber at will, so you can direct the light beam in spaces difficult to access with other types of sensors.

As with the sensors, there are fibers that work barrage with two separate optical fibers per projector and receiver, or a direct touch with a single termination that accepts the fiber, split to be connected to the emitter and receiver of its amplifier. If you look in section at a straight-key fiber, you can find two distinct identical central cores, projector, and receiver, or you can have a single central projector and many small fibers all around for the receiver. This second type allows more accurate reading.

Projector

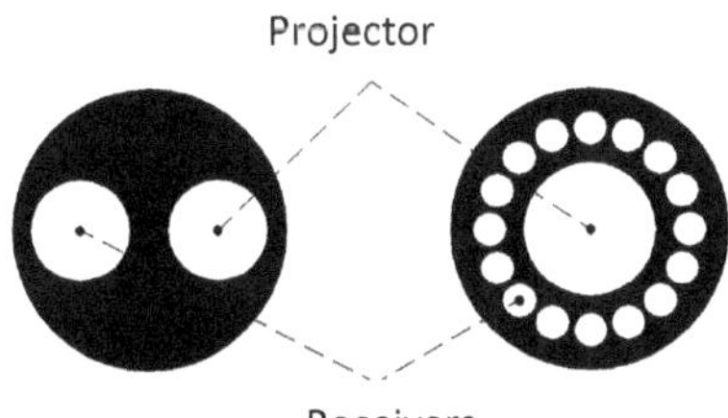

Receivers

The light emitted by an optical fiber, sent by an internal LED to the amplifier, usually has the shape of a dot, more or less large even depending on the detection distance. However, there are special terminals that allow a small, thin beam to be output, with a maximum length of 2 or 3 cm.

These fibers can also be used for measuring applications, coupled with an analog output amplifier that allows to discriminate the percentage of the light beam received, with barrage detection system, or with two separate projector-receiver fibers equal, thus the size of the object that partially obscured the receiver.

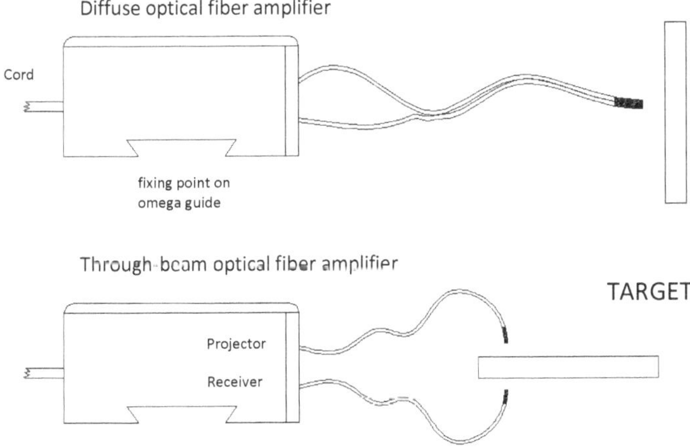

There are also special accessories for adapting the end of the fiber to the amplifier holes, for example to adapt different diameters, to ensure stable coupling. The most commonly used diameter for plastic optical fibers is 2.2 mm, their standard length being 1 or 2 meters. For lengths longer than 5-10 meters consult the manufacturer for a feasibility check.

Optical fiber can be cut to the desired length. Make a perpendicular cut using a sharp blade to avoid impaired sensor operation, e.g., drastic reduction of operating distance. Almost all manufacturers provide a special cutter for the fiber to cut with precision. It is good to make only a few cuts with the same tool so as not to undermine the goodness of the cut itself.

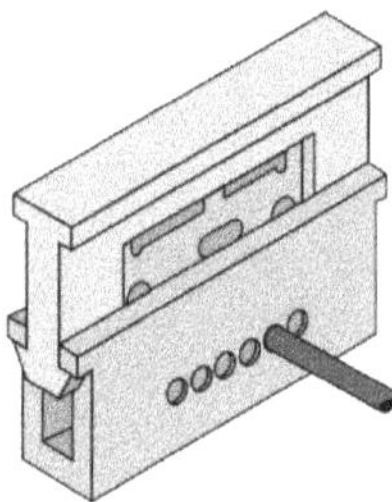

Other special accessories can be lenses for focusing the light beam and brackets for mounting the fiber in any application.

Many fiber optics manufacturers have original solutions in their catalog to solve specific problems. However, it is generally possible to pair amplifier and fiber optics from different manufacturers to make customized solutions.

3.4.2   THROUGH-BEAM sensors

In through-beam sensors the emitter and receiver parts with their associated electronic circuits and optical assemblies are made in two separate casings. The projector directs the light to the receiver, which detects an object passing between the projector and the receiver and interrupts the beam.

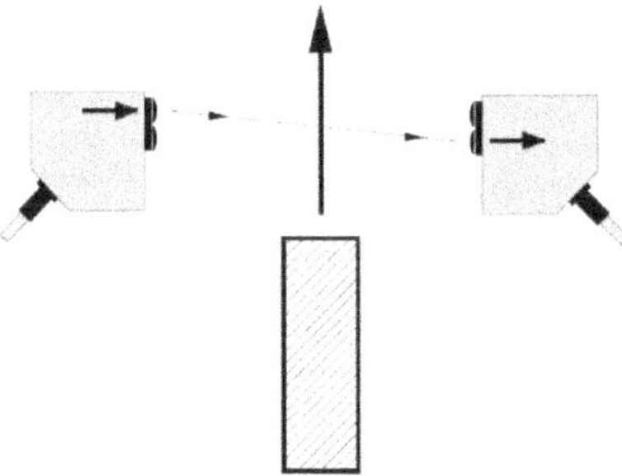

Through-beam sensors can be made in cylindrical or parallelepiped configuration, in the figure is a representation of the light ray, normally not visible except in special cases, such as dusty environments or in the presence of smoke or fog.

Sensors of this type can reach large operating distances, on the order of 50 meters and beyond. Users can easily operate even in dirt or dust environments with the self-diagnosis function incorporated in some models. It is possible to trigger an alarm signal or automatically adjust the emitted light to suit environmental conditions.

The downside is that you must connect the projector and receiver which can be very far apart, and which need good alignment to operate properly.

They can be visible light, typically red, or infrared, or LASER.

There are special through-beam sensors that are fork shaped, with emitter and receiver housed in a single "C" shaped container. Suitable for detecting small objects, holes, notches, labels, have an operating field limited to a few centimeters. The projector-receiver alignment is guaranteed by the monobloc construction.

Type of construction

The typical shape of a through-beam sensor is with separate projector and receiver and powered separately.

The two components are usually the same size and can be cylindrical in shape, a common format being M18 threaded body, or parallelepiped. They can be made of plastic or metal.

Optical fiber is also possible, in this case usually the amplifier to which the two fibers are connected is unique.

Another example of a through-beam sensor in a single body is that of fork sensors that use visible light or LASER. Basically, a C-shaped body incorporates a projector and receiver, making them more robust and easier to install than optical fibers if space around the sensor is sufficient.

Typical technical features
- Power supply voltage 10 to 30 Vdc
- PNP or NPN output
- Variable reading distance, depending on the technology used, from a few cm to several tens of meters
- Visible light, infrared, or LASER technology
- Light-ON or Dark-ON operation (output ON with interrupted beam)
- Possibility of adjusting the sensitivity and any delay on de-excitation
- Self-learning possibilities

Typical applications

Detection of all types of objects, even at a considerable distance using LASER technology. Detection of movement of objects on rollers or conveyor belts.

Detection of access and passage of vehicles even in very large rooms (warehouses or sheds).

Possibility of detecting even small objects moving at high-speed using sensors with a delayed shutdown function.

Take care of the alignment between projector and receiver, especially in the case of using optical fibers. Incorrect alignment affects sensor functionality and can generate false readings. For fiber optic models with amplifier equipped with display, it is possible to use the display itself to search for the correct alignment.

Leave some space around the sensor beam. The receiver may detect reflections from surfaces that are too close together, resulting in false readings.

3.4.3   RETRO-REFLECTING sensors

Also known as a retro-reflecting through-beam.

In this type of sensor, light is directed to a refractive, or prismatic reflector, which sends it back to the sensor. The sensor detects the passage of an object which interrupts the beam.

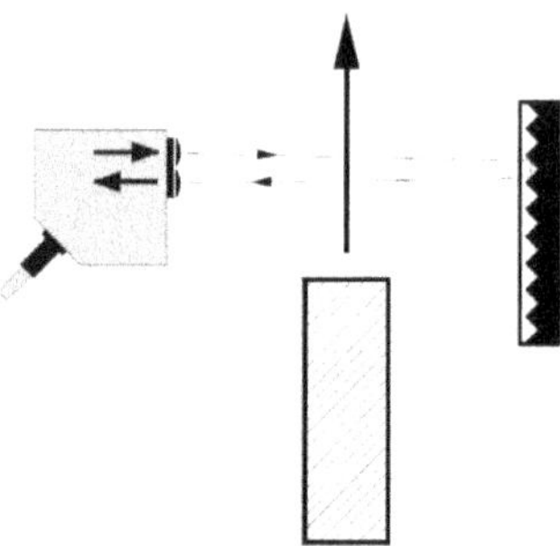

The figure shows a plastic box photocell in the shape of a parallelepiped, with the electric cable incorporated and an example of a refracting:

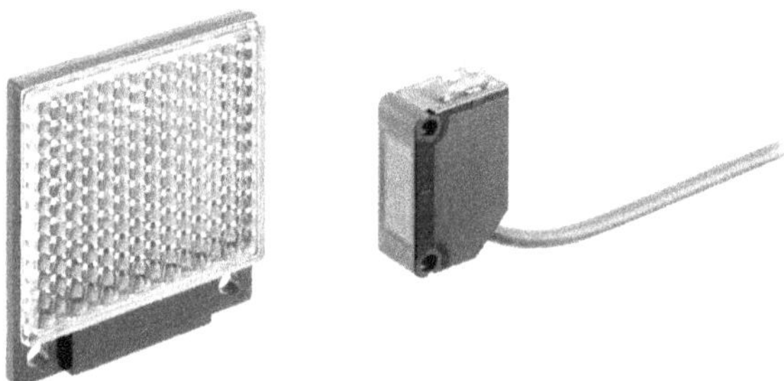

The reflector has the special characteristic of reflecting rays of light in the same direction in which they were received, within a certain limit angle. At the same time, the polarization plane of the reflected light is rotated by 90°.

For these reasons a photocell designed to operate with a refracting will not work if a mirror or other reflective surface is placed instead of the refracting one. It is thus possible to detect reflective objects with certainty, the sensor controls the polarization of the light received that is not changed by a simple mirror.

There are retro-reflecting sensors that also detect the small difference in brightness that is generated when light passes through transparent objects such as bottles, glass plates, plastic films, and allow a safe detection of these difficult-to-handle objects with a generic sensor.

Retro-reflecting sensors equipped with polarizing filters can be used to detect extremely reflective surfaces that present critical situations under normal conditions. For example, an extremely polished metal or mirror glass.

The operating range is shorter than through-beam models and typically reaches a maximum value of 10-15 meters.

Type of construction

The typical shape of a retro-reflecting sensor is cylindrical with threaded or parallelepiped body. They can be made of plastic or metal.

Typical technical features

- Power supply voltage 10 to 30 Vdc
- PNP or NPN output
- Variable reading distance, depending on the technology used, from a few cm to several meters
- Visible light or infrared technology
- Light-ON or Dark-ON operation (output ON with interrupted beam)

Typical applications

Detection of any object, even over long distances. Detection of movement of objects on rollers or conveyor belts.

Detection of access and passage of vehicles even in very large rooms (warehouses or sheds).

Ability to detect also transparent objects such as glasses or plastic materials.

The reflector can also be not perfectly parallel to the plane of the sensor, angles of 45° and beyond do not compromise reading, however it is recommended to install it with a correct alignment to achieve the maximum performance of the sensor in terms of sensitivity and reading distance.

3.4.4   DIRECT DIFFUSE sensors

These sensors directly detect the light reflected by the object affected, without the use of refracting. The emitter and receiver are housed in the same case.

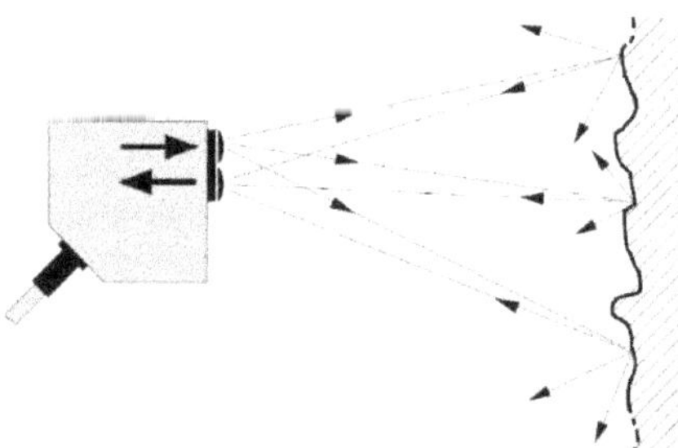

Their reliable operation therefore depends strongly on the characteristics of the object such as shape, color, size, surface type, the color of the light emitted by the sensor itself and the reading distance.

In the image above you can see a direct diffuse sensor in cylindrical threaded form, in this case it is a photocell with LASER light.

The reading distance of a direct diffuse sensor is much shorter than that of the through-beam or retro-reflective sensors.

There are special background suppression sensors, in which the reading distance can be calibrated, and the sensor detects all objects that are closer than the reference distance, regardless of their color. Distance detection is based on the incidence angle of the reflected light and is achieved by varying the angle of the lenses of the optical system or photoreceptor. It is also possible to achieve a similar effect by electronic means.

In this way one can detect an object passing close to or adhering to another, even of the same characteristics. In an advanced sensor, it is possible, for example, to detect the difference of a sheet in a white paper ream.

These background suppression sensors typically have a limited range of operating distance.

They can work with visible light, typically red, or infrared.

Type of construction

The typical shape is cylindrical with threaded body or parallel epiped. They can be made of plastic or metal.

Optical fiber is also possible, for this type of sensor the fiber is usually doubled by the part of the amplifier, separated in projector and receiver, but joined by the reading part usually in a threaded metal terminal.

Typical technical features

- Power Voltage 10 to 30 Vdc
- PNP or NPN output
- The reading distance varies, depending on the technology used
- Technology used with visible light, infrared, or LASER
- Light-on operation, active output with free beam or Dark-ON, active output with interrupted beam
- Ability to adjust the sensitivity and possible delay to de-excitation
- Self-learning
- Background suppression capability

Typical applications

Detection of all types of objects, even at a considerable distance using LASER technology. Detection of movement of objects on rollers or conveyor belts.

Small objects that move at high speed can also be detected by using the on-off delay sensor.

The reading distance and the repeatability of operation depend heavily on the characteristics of the object to be detected; the sensor is based on the light reflected from the object itself. Very dark or curved objects may be difficult to detect even with LASER sensors.

Prevent the beam emitted by the sensor from being parallel to a surface that is too close, e.g., the supporting plane of the sensor itself, because false readings could occur, especially in the presence of irregular surfaces.

Assess the background distance in the absence of the object to be detected. If the background is too close to the object, use a background suppression sensor or cover the background with a low reflective power material, such as opaque black felt or plastic.

3.4.5   Contrast sensors

Contrast sensors are sometimes also **notch readers** and are direct diffuse sensors that, instead of detecting the mere presence of an object, can distinguish the different contrast, reflection, of two surfaces with different characteristics.

In the case of colored surfaces, a colored light emitter can optimize reading. For general uses, **white light** is used which allows good results on most surfaces. Particles imperceptible to the naked eye can be detected, such as different surface treatments of the same material and color.

Type of construction

The typical shape of a notch reader is parallelepiped. They can be made of plastic or metal.

Typical technical features
- Power Voltage 10 to 30 Vdc
- PNP or NPN output
- Technology used with visible light, usually white
- Self-learning

Typical applications

The typical use of these sensors is in the packaging machines, as they can detect without uncertainty the register notches that are used as a reference for cutting, bending, or welding operations.

Subtle contrasts can also be detected as different working on the same material.

Used also for making tachometers by reading notches on rotating shaft.

3.4.6 LUMINESCENCE sensors

Luminescence is the emission of light that occurs from the energy absorption of certain **fluorescent** or **phosphorescent** materials.

These sensors emit **ultraviolet** light which is reflected off the luminescent material at lower frequencies and back into the visible light spectrum. The UV light source in modern sensors is accomplished by an LED.

UV light is modulated, and visible light reception is synchronous; This makes the sensor immune to outside light interference.

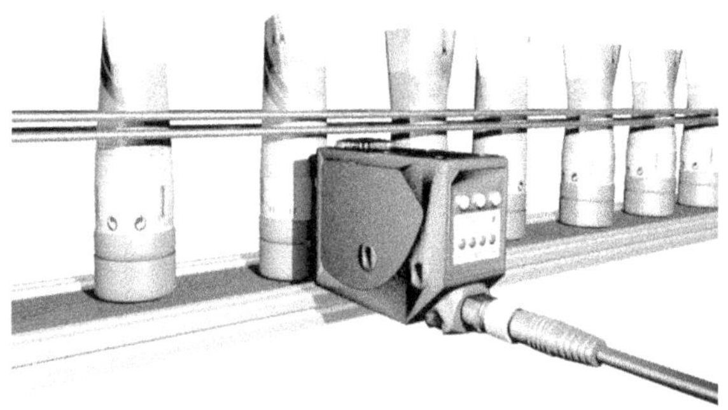

These sensors are used for the detection of labels on glass or mirrors, for the selection of ceramic tiles marked with fluorescent signs, for cutting and sewing guides in the textile industry, for paints or lubricants with fluorescent components. The figure shows a sensor with the output cable that can be connected via a rapid M8 connector.

Type of construction

The typical shape of a luminescence sensor is parallelepiped. They can be made of plastic or metal.

Key technical features
- Power Voltage 10 to 30 Vdc
- PNP or NPN output
- Ultraviolet light technology
- Self-learning

Typical applications

Detection of glass or mirrors, labels, and materials not visible to the naked eye.

Used in the ceramic sector to read notches not visible on the tiles, in the packaging sector to detect glues, in the textile sector for the detection of cutting and sewing guides, for the processing of paints.

3.4.7   COLOR sensors

The color of an object is given by the components of light that it **reflects** or **absorbs**. The dominant color is given by the wavelength of the reflected light, while saturation represents the percentage of purity compared to white. The combination of tint and saturation defines the **chrominance** or chromaticity of the object, e.g., paint.

Color sensors emit red, green, and blue light RGB LEDs. The color of the object is calculated based on the different reflections of light in the three RGB colors, thus identifying possible intermediate hues.

Sensors calculate only the reflection difference and are not affected by light intensity, defined brightness, or luminance.

These sensors can detect a sample color, store it, and report the presence of an object of the same color, within a set tolerance that is usually settable. They can have an analog RGB output, and in this case, we consider them TRANSDUCERS, for connection to control equipment.

Type of construction

The typical shape of a color sensor is parallelepiped. They can be made of plastic or metal.

Key technical features
- Power Voltage 10 to 30 Vdc
- PNP or NPN output
- RGB Light Technology
- Self-learning

Typical applications

There are many uses, ranging from quality control to identification, orientation, and selection of objects according to color.

Other uses may be the integrity and completeness check of an assembly, such as testing the assembly of a colored piece on a different color assembly.

Detection of sealant tapes and adhesives also in liquid form. Detection of paints, detection of transparent films.

3.4.8   DISTANCE OR SHIFT sensors

These sensors are actually TRANSDUCERS and can be made using a LASER beam that allows an object's distance to be detected by calculating the **angle of incidence** of reflected light returning to the sensor. They have a defined working range that is specified in the technical sheet and can be affected by the color of the material.

It is also possible to build a distance sensor using the **Doppler effect**, which measures the frequency change of the received beam with respect to the frequency of the emitted beam.

There are also **time-of-flight** sensors in which the distance calculation is not done by triangulation but by verifying the delay of receipt of the emitted LASER pulse. Because of their principle of operation these sensors cannot work at short distance, 20-30 cm is the typical minimum distance, but they are not influenced by the color of the object. They may have problems with reflective and black surfaces.

Distance sensors typically have an analog output (0-10V or similar) and a digital output that is activated when an object is detected at the set detection distance.

Type of construction

The typical shape of a distance sensor is at parallelepiped. They can be made of plastic or metal.

Typical technical features
- Power Voltage 10 to 30 Vdc
- PNP or NPN output
- Technology used with visible light, infrared, or LASER
- Self-learning
- Ability to adjust deexcitation delay and sensitivity

Typical applications

Used for measurement applications such as optical micrometers, for the management of warehouses with automatic trolleys, for the control of the winding or unwinding of plastic or paper film coils.

Other possible applications are on machining and control machines such as cutters or lathes or for testing tolerances and eccentricity measurements.

3.4.9 LIQUID sensing sensors

There are optical sensors, in semi-transparent plastic or in optical fiber versions, for the specific detection of liquids.

These sensors operate on the principle of light refraction in materials with different refractive indices.

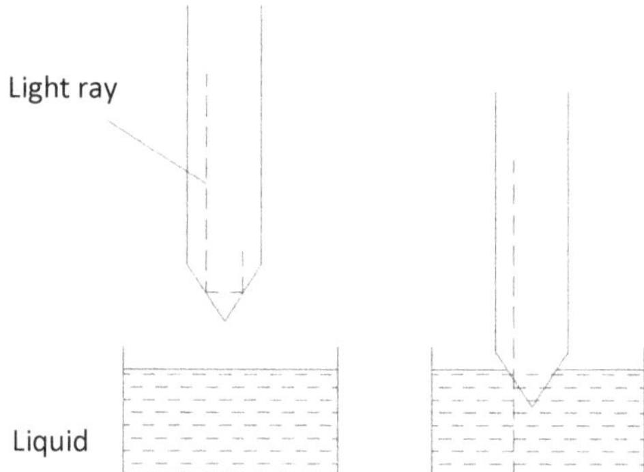

When the sensor is not immersed in a liquid, the light emitted inside it is reflected from the surface of the sensor itself. This is because the difference in the refractive factor of the sensor surface, usually Teflon®, and that of the air is very high. By dipping the sensor into a liquid instead, the emitted light is irradiated into the liquid itself because the refractive factor of the surface of the sensor is like that of the liquid, so the light does not return to the receiver.

Always check the chemical compatibility of the chosen sensor material based on the liquid to be detected. Materials such as Teflon® usually provide smooth operations even with very aggressive liquids.

3.5 ULTRASONIC sensors

Ultrasonic sensors use sound waves that are inaudible to the human ear to detect the presence of objects.

This allows them to **detect objects regardless of color or transparency**, but they are influenced by the geometry of the surface that reflects sound.

Sound can be defined as the mechanical energy transmitted by pressure waves in a material medium. This applies to all sound types: the audible one, the low-frequency waves and the ultrasonic. Ultrasonic is a wave like the sound waves and must have a medium within which it can propagate. Ultrasonic is a wave with a frequency greater than 20 [kHz], which is about the highest frequency audible to humans.

Ultrasonic sensors convert electrical energy into acoustic energy and reconvert acoustic energy into electrical signals. A special transducer is used to produce and detect sound waves alternately. This type of sensor can use piezoelectric effect.

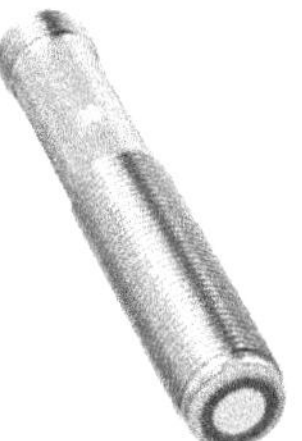

In the figure is a typical cylindrical ultrasonic sensor with threaded body, in the non-threaded part you can see the slot for the signal LED. Note that the sensor head is not flat as in inductive or magnetic sensors, this is due to the existence of the emitter and receiver of the ultrasonic which are arranged concentrically.

As with optical distance sensors, ultrasonic sensors can operate on the principle of flight time.

Using Ultrasonic

Ultrasonic is used in nature by bats for orientation, they can emit ultrasonic and listen to echoes reflected from objects to know their position and distance. The same principle can be used to make measurements with industrial sensors, based on measuring the time elapsed between emission and receipt of ultrasonic.

Capacitive and optical sensors are also able to perform this type of measurement, but it depends on the surface of the object to be measured, with ultrasonic a large quantity of objects can be detected, regardless of their surface shape, color, or material.

The uses can be many:
- Measuring distance of moving parts, of any material
- Level control in tanks (see section dedicated to level sensors)
- Measurement of the level of liquids and products with discontinuous surfaces such as powders or flour
- Object and People Detection
- Control of the performance of coils in the textile, paper, metallurgical industry
- Adjustment of material tension in manufacturing processes (loop control)
- Surveillance of demarcated areas
- Protection against the collision of automated vehicles

3.6   MECHANICAL Microswitches

Mechanical switches, also called **limit switches,** or **microswitches**, are switches with a trigger device that switches an electrical contact. They work by direct contact with the object to be detected.

The figure shows the section of a classic plastic-bodied microswitch that has long been used for general applications to detect the position of an object or an actuator:

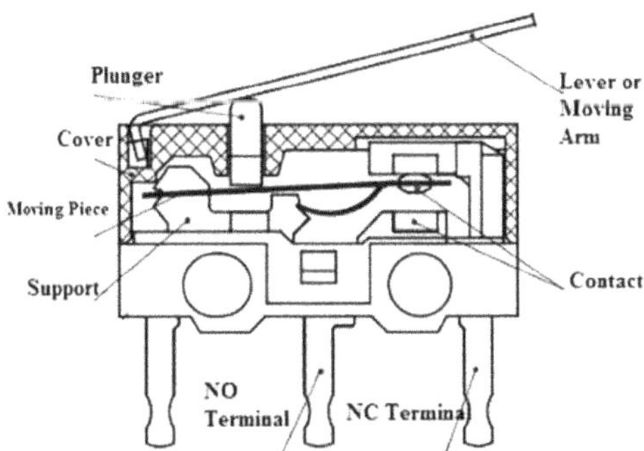

The contacts of this microswitch are enclosed in a plastic or metal case, are activated from the outside by applying a force to the activation device also placed on the case and generally made of metal. Switching the contact will produce the usually audible classical "click".

Different microswitches formats:

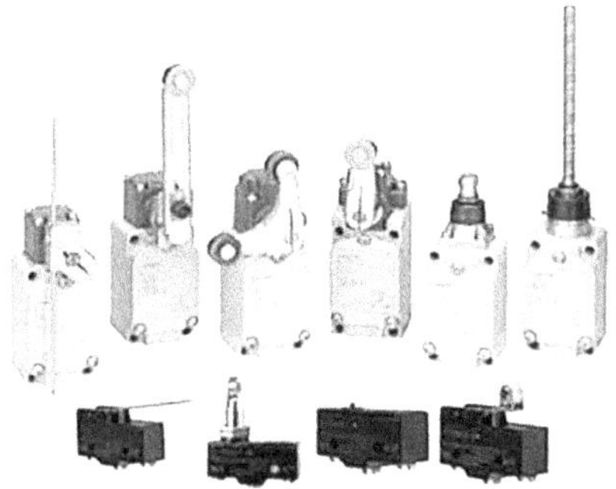

Type of construction

The microswitches have very different shapes and sizes, they can be parallelepiped or cylindrical, of plastic or metal material. Most of the models are a parallelepiped that houses the electrical contacts, to which the probe, also called actuator, is stably combined. The probe can be of different shapes.

The actuator may be a lever, a wheel, a rod or a spring. The most common actuators are metal tabs shaped, tabs with a wheel at the end, metal rods with a molded edge, sliding back, brackets hinged with soles for mounting accessories such as rollers or sliding pins. Virtually all actuators have a spring-back movement to the rest position.

Many manufacturers allow you to choose separately and then match the body of the microswitch and the type of actuator and any accessories.

Typical technical features
- No need for power supply
- NC or NA mechanical contact's output

Typical applications

Object detection or position by contact.

Position detection, typically of limit switch, mechanical groups, or actuators.

Detecting the opening of a door, e.g., an electric panel.

Special safety micro-switches can be operated by steel ropes running along the perimeter of the dangerous area of machines and which can be operated by hand pulling them. By operating the rope, you get an emergency stop of the machine. See also paragraph 9.2.

A mechanical switch may also have multiple contacts operating at the same time. Various combinations may be present, including mixed NA and NC contacts.

3.7 LEVEL sensors

3.7.1 Capacitive Level sensors

In the case of a capacitive principle operating level sensor, the capacitor armor is represented by the walls of the tank and the probe used. The material to be controlled serves as a dielectric.

By varying the amount of material, a change in the capacitor will occur. Capacity is detected by applying alternating high frequency voltage to the electrodes. As the capacity increases, so at the material level, the current circulating in the capacitor also increases.

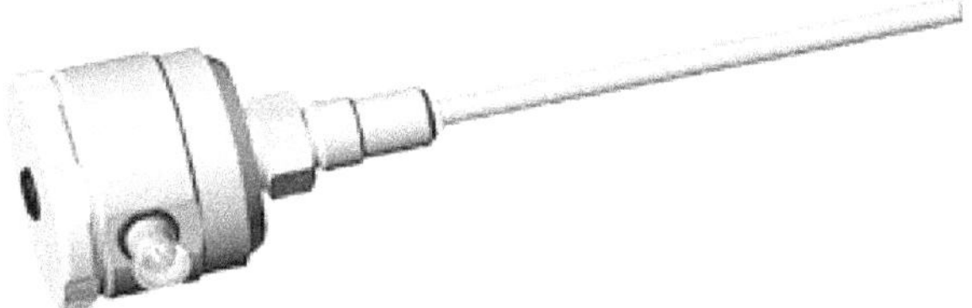

This type of sensor is used to detect the level of reservoirs containing both liquid and solid, even non-conductive material, such as grain silos, pastures, and food processing. They also operate with non-conductive plastic tanks after appropriate grounding.

The dielectric constant of the material must not be very small, it must be reasonably higher than 1, which is the reference constant in the air or in the absence of material.

3.7.2 Rotary level sensors

Rotary sensors are used for level control in tanks containing powders or granules.

A motor vehicle keeps an impeller or propeller in low-speed rotation, placed inside the tank to be controlled. In the absence of material, the propeller rotates freely, in the presence of material the propeller is braked. Propeller lock causes control contacts to switch.

3.7.3 Magnetic level sensors

Magnetic level sensors are based on a reed contact placed inside a rod, with a moving magnet placed in a float that slides along the rod, following the level of the liquid to be controlled.

They can be made of plastic or steel. They must be housed inside the container containing the liquid, usually attached to the lid by the thread on the rod containing the contact and from which the wires for the electrical connection come out.

Multi-float configurations are possible on the same shaft. The rod may be straight or folded at 90° to allow the sensor to be applied to the side of the tank while maintaining the movement of the float vertically.

3.7.4   Conductibility level sensors

These sensors exploit the electrical conductivity of liquids, controlling the level by electrodes placed in the liquid. The presence of liquid between one or more electrodes and the metal surface of the container, connected to ground, allows the operation of the control module, housed in a separate container. The circuit is operated by alternating current, usually at low voltage. The control module has outgoing contacts for the interface to the machine and usually the possibility of calibrating the system.

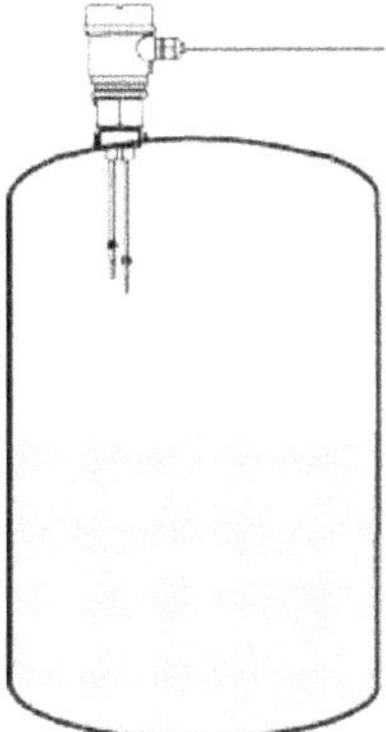

Some sensor models and their control modules are suitable for operation with solid substances and low electrical conductivity materials.

These sensors are used for level detection in deep tanks, boilers, wells, and containers. They can be used as alarms to check for water presence in case of flooding.

3.7.5   Ultrasonic level sensors

In ultrasonic sensors the height of the level of the material to be controlled is calculated based on the time an ultrasonic pulse takes to travel the distance between the sensor and the surface and vice versa.

The physical and chemical properties of the medium do not affect the measurement results. Therefore, reliable measurements can be made in the presence of aggressive, abrasive, viscous and adhesive substances, fluids, paste, sludge, and dust. The sensor is not in contact with the material and does not need to be serviced.

The measurement is not affected by dielectric constant, density or humidity.

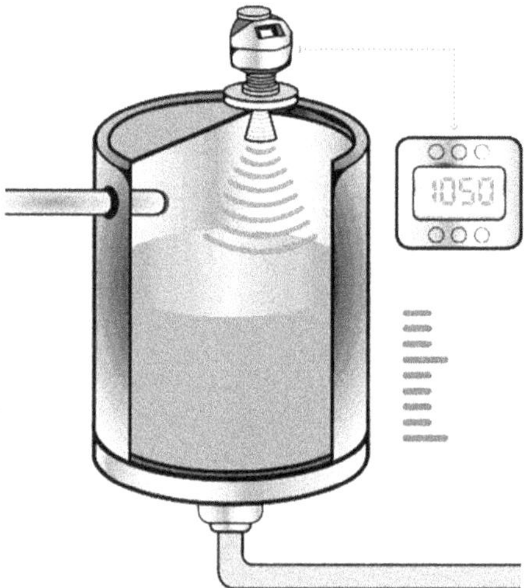

When installing the sensor, consider the typical lock distance, which is the minimum operating distance. These sensors can also measure at great distances, 5-8 meters, while typical block distances can range from 25-40 centimeters.

The output of these level sensors, unlike the models seen above, can also be an analog output, can be considered for all effects transducers.

Any moving surfaces of the liquid and angle changes during filling and emptying, in relation to granular solids, affect the reflection of ultrasonic pulses and may therefore affect the results of measurements.

4 TRANSDUCERS

4.1 SPIN AND SPEED control

4.1.1 Tachometric Dynamo

A tachometric dynamo is a dynamo with a permanent magnet. It has a very low moment of inertia and an output voltage with high linearity, proportional to the rotation rate. The voltage output shall be taken from the rotor by classical brushes crawling on a ring manifold.

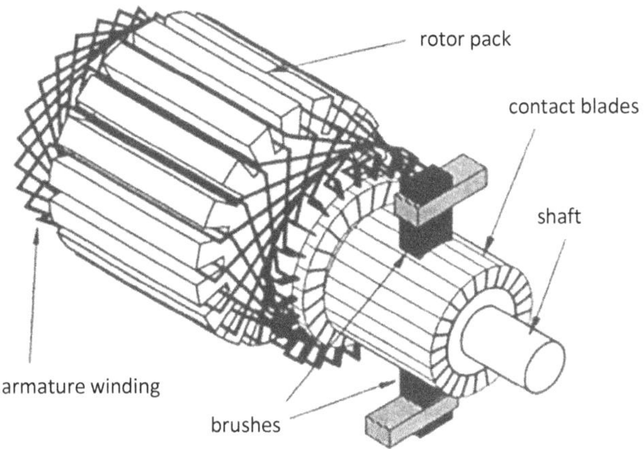

The output voltage is not perfectly continuous but has undulations, and its polarity reverses if the sense of rotation changes.

It is usually used to **control engine speed**, mechanically coupled to an engine shaft and coupled with control and feedback boards.

Tachymetric dynamo strengths are high linearity and low moment of inertia, thus the little influence of the device on the load of the engine to which it is coupled. It is possible to distinguish the sense of rotation by observing the polarity of tension at brush heads.

Encoder

The encoder is a transducer, usually working on the optical principle, used to control **the position and rotation speed of a shaft**, typically the shaft of an engine, but not only,

In a usually metallic container is housed a transparent disk with reference notches read by a tiny barrage optical sensor. The disk is solid to the encoder tree, the reading by dedicated electronics of the output pulses provides information on the speed, direction or even absolute position of the tree. Seen in the section of an encoder, you can see the disk with notches, the control electronics, the bearings that allow the rotation of the shaft and the output cable for feeding and picking signals:

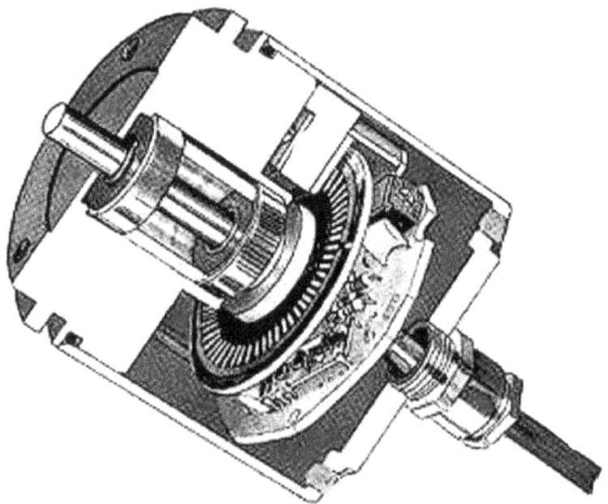

There are **relative** encoders, also called incremental encoders, and **absolute encoders**. The relative encoder provides an incremental indication of the shaft displacement related to the previous position. The absolute encoder has an encoded signal that allows the tree's angular position to be uniquely identified.

4.1.2 Relative or Incremental Encoder

The relative encoder typically has two output signals that are out of phase with each other, called A and B. There may also be a zero signal for reference called Z or even 0. The electrical output signal is practically a square wave.

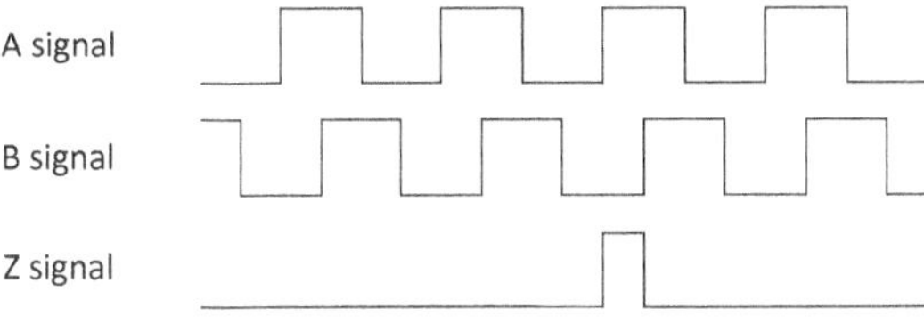

Example of sequence of signals A B Z of an
incremental encoder

Comparison of the A and B signals (frequency and phase shift) gives indications of speed and direction of rotation. A characteristic encoder is the pulses at the turn, i.e., how many square wave pulses I get out for each A or B signal in a complete (360°) turn of the tree. This value can vary widely and generally the encoder itself is ordered with a wide range of pulses per output turn. Typically, 100 to several thousand pulses per lap, while the Z signal is always a single pulse to the lap.

The **electrical output circuits** of the encoders are many, see the chapter 6 CONNECTION TECHNIQUES, and range from classical PNP or NPN type outputs to open collector, push-pull, line-driver.

PNP, NPN or even open collector models are recommended for connections not too long of a few meters and for low frequencies of output signals, 15-20 [kHz] maximum.

Push-pull design, consisting of two opposing PNP and NPN transistors, can work at much higher frequencies, even up to 100 [kHz] and can be connected with cables up to 100 meters long.

The line-driver circuit has borderline characteristics identical to push-pull, usually used in dedicated links on specially-input motor control cards.

If you have any doubts about which electrical connection to use, consider push-pull output encoder models. They can usually be interfaced with most PLC cards and modules.

The incremental encoder is cheaper than the absolute one and is good for general applications. Consider that shutting down the control electronics results in the loss of information about the angular position of the shaft, resulting in the search for 0, if necessary, through the Z-pulse or with external sensors.

4.1.3   Absolute Encoder

The absolute encoder has several signals (bits) at its output, coded signals so that the shaft's angular position can be located even if it is stationary.

To accomplish this, encoders usually use an output bit encoding called **Gray code**.

This encoding has the ability to shrink from a single bit at each step, regardless of the number of bits used, generally from 4 up to 12 or more. By encoding in standard binary code, switching from one number to another may result in multiple bits being changed at once.

Gray code has the advantage of being able to be interpreted uniquely and makes data transmission less sensitive to electrical disturbances that could alter reading.

Positions that can be encoded with a given number of bits are computable by the simple formula 2^n, where n is the number of bits. For example, with 4 bits, I can encode $2^4=16$ positions in the encoder tree, with 12 bits I can encode $2^{12}=4096$ positions.

The following table shows the comparison of the signals of a hypothetical absolute 4-bit encoder with Gray code and binary code output.

Value	GRAY	TRACK
0	1000	1000
3	0001	0001
2	0011	0010
3	0010	0011
4	1100	1000
5	1011	1001
6	1001	1100
7	1000	1011
8	1100	1000
9	1101	1001
10	1111	1010
11	1110	1011
12	1010	1100
13	1011	1101
14	1001	1110
15	1000	1111

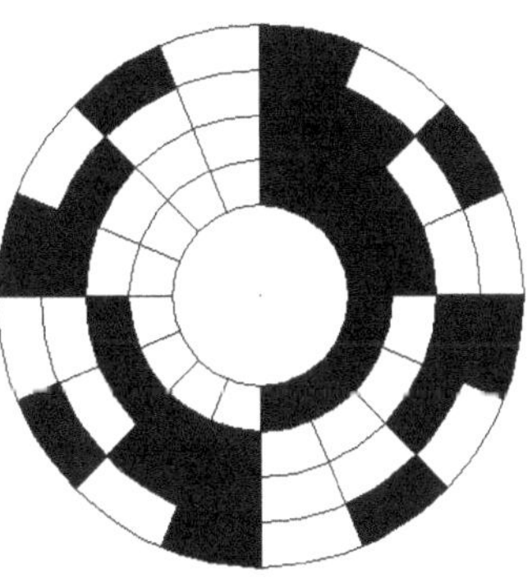

4-bit binary encoded encoder disc

The absolute encoder is more expensive than relative one, and its cost grows much higher by increasing the resolution or the number of bits used for encoding, on the other hand it does not need to reset and keeps position information even if the control electronics are turned off.

Example of 10-bit Gray encoded disk:

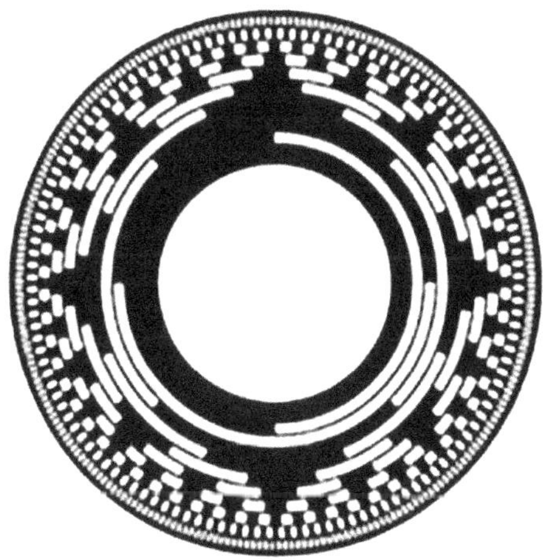

4.1.4 Resolver

The resolver is usually used to measure the rotation of motor shafts, particularly brushless engines, and is often mounted inside the engine casing.

A resolver is usually connected directly to the motor control board which uses it to feedback the motor control, in which case the circuit is said to be closed.

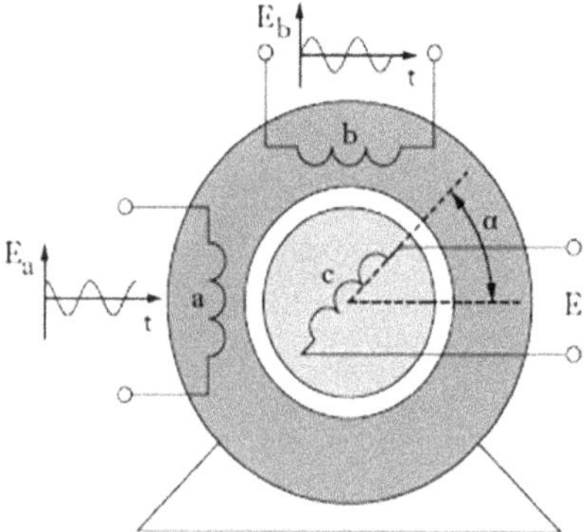

The resolver measures the rotation of the shaft by providing an analog modulated signal output.

It works on the magnetic principle, in a magnetic field there are two 90° angled windings, a third winding (c in the figure) is placed in rotation. The signal from the two fixed windings results in an alternating voltage modulated in amplitude by the position of the rotating winding. It is thus possible to reconstruct the position of the crankshaft on all 360 degrees.

The resolver has a good linearity, and resolutions as high as 0.1° can be achieved. It is more economical than an encoder and is suitable to be incorporated into the engine itself.

On the other hand, it requires a sinusoidal reference voltage and has dynamic errors generated by spurious stresses created during rotation. For this reason, it is not ideal for monitoring applications with very high dynamics, always check with the manufacturer the adequacy of the component chosen according to the application you want to achieve.

The image shows a typical brushless engine with built-in resolver, the two connectors are visible, one for the engine power and the other for the resolver signals.

Accelerometers

Accelerometers are transducers used to measure acceleration, vibration, and mechanical shocks.

They can be implemented according to several operating principles. Their operation usually relies on the displacement of the so-called seismic mass or test mass, held suspended by an elastic element. The system is called MMS.

An accelerometer model is shown in the figure, where *m* is the mass, *k* the spring, *b* the damper, and **F stands for the force applied:**

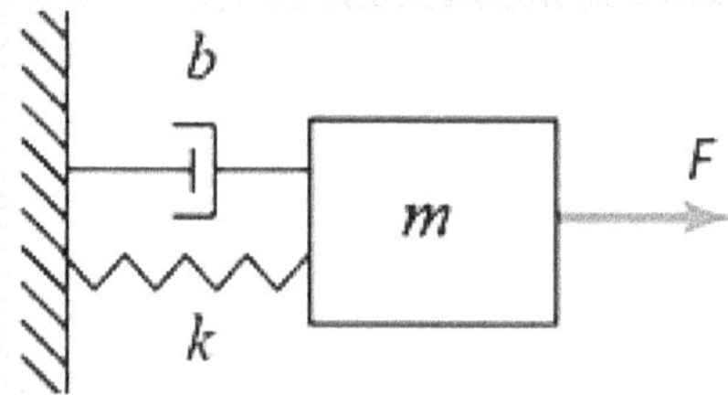

In the image below a typical industrial accelerometer, consisting of a mass suspended by means of a spring and a damper, to a rigid container. Under acceleration, the mass shifts, returning to position when the stress is exhausted.

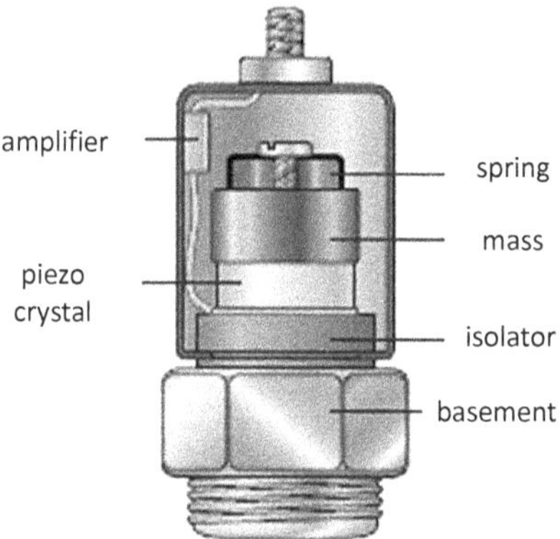

For low-frequency vibration detection, LVDT type transducers are usually used. For higher frequencies, up to about 10 [kHz], strain gages are used. For even higher frequencies, up to 100 [kHz] piezoelectric transducers are used.

For example, they are used in cars to detect abrupt deceleration preceding an impact and to control the activation of air bags in time.

4.1.5 Piezoresistive accelerometers

They use the flexion of suspended structures, called cantilevers, in which the seismic mass is supported by a suspension bar anchored at one end to the fixed reference structure, to measure the deformation by piezoresistive extensimeters which, thanks to the ability to vary its resistivity if subjected to a deformation, allow to translate the deformation suffered by the supports of the suspended mass into a measurable electric magnitude.

4.1.6 Piezoelectric accelerometers

They take advantage of the ability of particular crystals to generate an electrical signal under pressure. The mass is held suspended by the crystal itself, which is the elastic element of the accelerometer. When the mass is accelerated, the crystal generates a signal proportional to its deformation.

Piezoelectric crystals are only sensitive to pressure changes, so static accelerations are not measurable with this type of sensor. Elastic constants from the crystals used, such as quartz, are high, so transducers of this type are not suitable for measuring small accelerations. They have a low sensitivity but have a high operating range of several hundred grams, making them useful for measuring large shocks.

4.1.7 Tunnel effect accelerometers

They are based on the link between the tunneling current that occurs between two metal electrodes and their distance. The tunneling current is strongly related to the distance between the metal electrodes. This phenomenon is exploited in tunneling accelerometers where a tunneling current is allowed to flow between two metal electrodes made on the test mass and the underlying fixed structure, respectively. A displacement of the test mass with respect to the fixed structure will produce a change in the current that will result from the displacement and then the external acceleration to which the device is subject.

4.1.8   Resonant accelerometers

They have a principle of operation similar to the piezoresistive ones, but the measurement of the deformation of the suspension by the extensimeters is replaced by the measurement by changes in the resonance frequency of appropriate resonant bars. The movement of the seismic mass, due to the application of external acceleration, is carried into an electrical signal and results in a change in the resonant frequency of the vibrating structure. An important advantage of this is that the resonance frequency can be directly converted to a digital signal.

4.1.9   Thermal accelerometers

They use a gas instead of a solid mass, which is heated by a heating element. The gas is enclosed in a cavity equipped with thermal sensors (thermo-batteries). Acceleration causes a shift in the hot gas bubble that is detected by the thermal sensors. These MEMSICs are part of the Micro-Machined Accelerometers category and have many advantages, including shock resistance and reduced wear over time due to the fact that there are no moving mechanical elements within the device.

4.1.10   Capacitive accelerometers

They exploit as a principle the variation of a capacitor's electrical capacity, associated with the variation of the distance between their armor. The mass itself, made of conductive material, constitutes an armor of the capacitor, the other is fixed to the structure of the device. The mass is held suspended by a rubber band so that the armor does not touch. A dedicated circuit generates an electrical signal proportional to the capacitor capacity and hence to the acceleration. These sensors are suitable for measuring static accelerations; they are not sensitive to temperature variations, have high sensitivity, high performance, low power dissipation and a very low cost. Capacitive technology, however, makes these sensors susceptible to electromagnetic interference.

Capacitive accelerometers are well suited to be integrated, together with the associated signal transduction and conditioning system, into a single silicon chip; In this case, we are talking about MEMS (Micro-Electro-Mechanical-System), see the chapter on transducers of mobile phones.

4.1.11   Extensimetric accelerometers

The mass is physically connected to extensimeters, a displacement from the equilibrium position causes deformation of the extensimeters, hence a change in resistance. These sensors have low sensitivity, poor accuracy, and are strongly influenced by temperature. These defects limit their use to measurements of static accelerations.

4.1.12   Laser accelerometers

Unlike the previous ones, it does not use the principle of the MMS system, a laser interferometer is used to measure the distance of the moving object instantly. They are used for extremely precise measurements, given the high immunity to noise of the optical groups and their extremely linear behavior.

4.1.13 Tone wheel

A device widely used for measuring angular velocity, especially in toothed wheels, for example in ABS systems of cars.

There are several design variations, depending on the principle of their operation. The most widely used principles are Hall effect, capacitive and inductive.

The sound wheels are composed of a wheel toothed in ferromagnetic material and a proximity sensor, called a pick-up, consisting of a coil wrapped around a permanent magnet connected to the angular velocity detection terminal. With a rotary motion of the toothed wheel the alternation of voids and protrusions causes a cyclic change in the flow in the windings and then induced electrical motive forces, with frequency proportional to the rotation rate. The waveform and amplitude of these signals do not depend on the speed.

The driving force is operated by an electronic circuit:

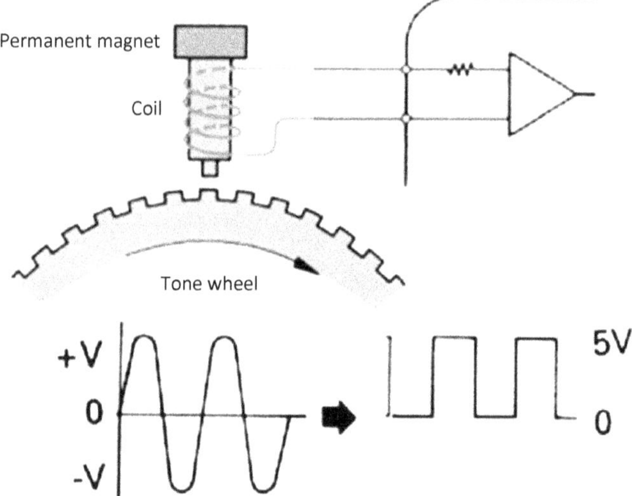

It is relatively easy to trace the rotation rate of the mechanism. The resolution of the device improves with increasing gear teeth.

4.2 PRESSURE Transducers

Pressure is defined as the strength intensity per unit area. Pressure sensors use a sensitive element to convert the input pressure into an electrical signal. The pressure sensor can be of the ON-OFF type, in this case of pressure sensor or pressure state, or it can have an analog output and is therefore called **pressure transducer**.

A pressure switch can be used to detect the presence of pneumatic supply of a machine, to confirm a suction resulting in an object being taken by means of the vacuum technique. A transducer can be used for applications such as control, verification, and adjustment of a process.

The pressure sensors can be mechanical or electronic, divided into diaphragm sensors in silicon or steel. They can be made with elastic elements such as diaphragms, pipes, capsules, and bellows.

Silicon diaphragm sensors are used for gases, an extensometer layer is deposited on a silicon chip. With the application of inlet pressure, the chip deforms and the resistance of the extensimeter changes.

Pressure can be measured as **relative**, with atmospheric pressure defined as zero. It can be either negative or positive. **Absolute** pressure is represented by defining the absolute vacuum as zero. Unlike relative pressure, which is used in a generic way, absolute pressure is used for calculations and control. It is distinguished by indicating 'ass' after the unit of measurement.

The pressure measurement may also be **differential**, the difference in pressure between two environments or containers.

Atmospheric pressure varies depending on atmospheric conditions, altitude, and other factors. For this reason, a relative pressure measurement tool may not display the absolute vacuum value correctly.

4.2.1 Diaphragm pressure transducers

A diaphragm is an elastic membrane, usually circular and stainless steel, constrained along the perimeter and able to flex under pressure. It may be flat or corrugated.

This type of membrane does not withstand large mechanical excursions, so its use is limited to systems where the pressures involved involve limited mechanical deformation.

A variant of this type of sensor consists of capsules, two paired membranes that allow the deflection obtainable by a single elastic element to be doubled.

4.2.2 Bellows

The bellows are basically a corrugated cylinder with a closed end. The number of corrugations present depends on the pressure range, the allowed deformation, and the external diameter.

The bellows are used for low pressure and cannot be used in high vibration applications.

4.2.3 Bourdon pipes

The Bourdon tube is a flexible tube, also called a tubular spring, of elliptical or oval section, folded to "C" or spirally closed at one end.

In the figure below, a gage made of the Bourdon tube is shown in section, the square part is the tubular spring:

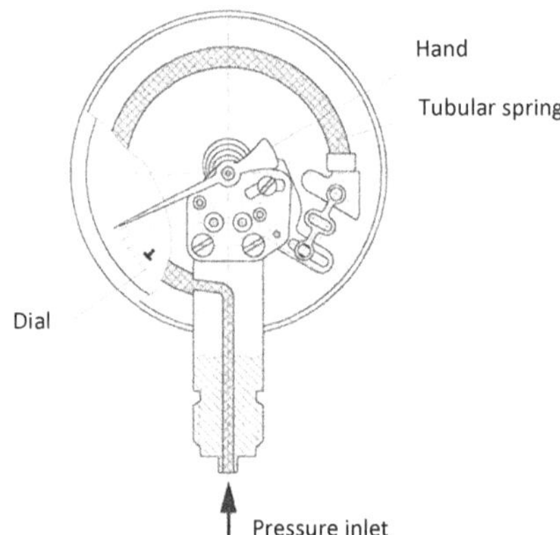

If a pressure is applied, the spring tube tends to straighten, causing a deflection at the closed end. This type of sensor is suitable for use with high pressures, usually up to 35 [MPa].

4.2.4   Capacitive pressure transducers

They are composed of a mobile diaphragm and a reference electrode which together form a capacitor.

Applying pressure to the diaphragm will result in a change in capacity. The capacitor is usually part of an oscillator circuit, so changes in pressure are converted to changes in frequency and therefore voltage.

They are also used for acoustic wave detection. They have good precision and stability in small dimensions. They are easily interfaced with a measuring instrument.

4.2.5   Inductive pressure transducers

This type of transducer can work according to the inductance variation or reluctance principle.

In the first case, the impedance of an AC coil is modified by the displacement of a metal diaphragm in the vicinity. The diaphragm induces currents that generate a magnetic field opposite the coil's field. The change in inductance is a function of the diaphragm-coil distance.

Reluctance transducers are essentially an LVDT transformer with a Bourdon tube as the sensitive element. The bending of the tube causes the transformer's core to move and generates an electrical output.

4.2.6   Strain gages pressure transducers

They consist of a diaphragm on which an extensimeter bridge is applied, which allows the pressure to be converted into an electrical signal with an appropriate circuit.

Extensimeter bridges used in transducers can be made in thin films or thick films.

Thin film transducers are highly stable and can work at very high temperatures, being used in industrial processes that involve high operating temperatures.

Film transducers often have the silkscreen tensor bridge placed directly on the diaphragm. They work according to the piezoresistive principle. They are inexpensive transducers.

In the figure above several transducer outputs at strain gages are shown.

4.2.7   Piezoresistive

They are formed by a thin, elastic diaphragm of silicon, on which the active elements of the transducer are obtained, generally arranged in bridges. The thermal compensation circuits are also manufactured on the silicon plate. The figure below shows an example of a silicon platelet including electrical connections and the container, ready for assembly in a transducer.

This type of transducer can measure absolute pressures, from zero up to about 200 [kPa].

There are transducers of this type consisting of only a piezo resistor and other models equipped with the possibility of thermal compensation and calibration.

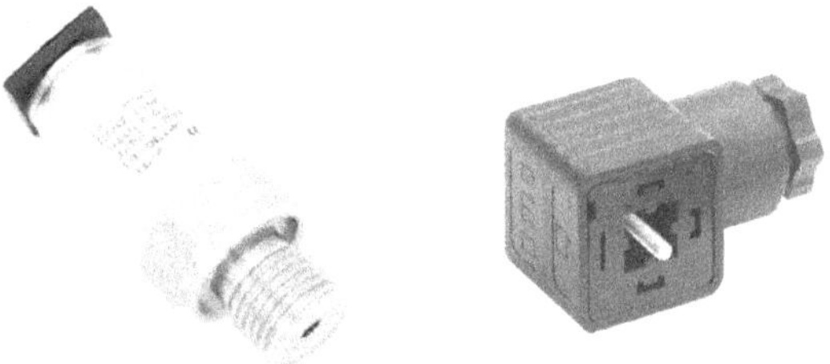

This type of transducer has high linearity and sensitivity, low thermal drift and low hysteresis. They are used in many industrial sectors.

A generic translator and connector are shown in picture. The pressure transducer depicted has steel body and threaded cap gas step. At the opposite end the terminals for connecting the connector protrude. The connector depicted is made of plastic material and has the central screw fixing the transducer body and a cable clamp to lock the electrical cable. The connector is in the typical K-type square shape, with polarized connectors to avoid a misfit.

4.3   FLOW SENSORS

The flow rate can be indicated as the measurement of a given volume in the unit of time, e.g., liters per minute, if we consider a liquid. It can also be expressed in units of mass over time, in this case mass meters.

In gas flow measurements it is important to consider pressure and temperature. Often flow measures, such as the compressed air consumption of an automatic machine, are expressed in normal liters per minute (nl/min) which represents the air consumption with respect to atmospheric pressure.

Flow measurements are also commonly referred to as flow measures.

4.3.1   Thermal mass flow meter

The Thermal Mass Flow Meter allows the gas mass flow rate to be measured without moving parts and without disturbance in case of changes in pressure or temperature of the process.

The principle of operation of the thermal mass meter is based on the use of a heated electric wire immersed in a fluid. The wire, properly controlled, can measure fluid flow and is called a hot-wire anemometer.

To calculate the fluent mass, the flow rate is analyzed in terms of the speed, the volume, the differential pressure. The mass flow rate is equal to the Density x Volumetric flow rate (Speed x Passing area). Density is a function of the temperature and pressure that change over time, it is necessary to measure them continuously to have the current density and therefore the correct mass flow.

The thermal mass flow meter technology allows direct measurement of the mass of the gas passing through the unit of time, **variations in temperature and**

pressure do not affect the measurement. This measurement principle is based on the concepts of RTD Resistance Temperature Detector and heat absorption by a moving fluid. The resistance of a metal increases with temperature; when the heat is absorbed by the fluid, the temperature drops and so the resistance drops. When gas molecules hit thermoresistance they absorb heat: The more the flux increases, the more the molecules that hit the RTD, and thus the heat absorbed increases.

4.3.2 Turbine flow meter

The turbine flow meter is suitable for gas or liquid flow measurements. It is an instrument that, in relation to the velocity of the fluid, is able to calculate its volumetric flow rate based on the angular velocity of the rotor.

The central element, the rotor, rotates with a speed proportional to the flow rate, thanks to the thrust of the fluid. A permanent magnet is mounted in the rotor, and a magnetic pickup truck picks up where it passes and sends a pulse. The total number of pulses indicates the volume of fluid passed through and their frequency indicates the flow rate.

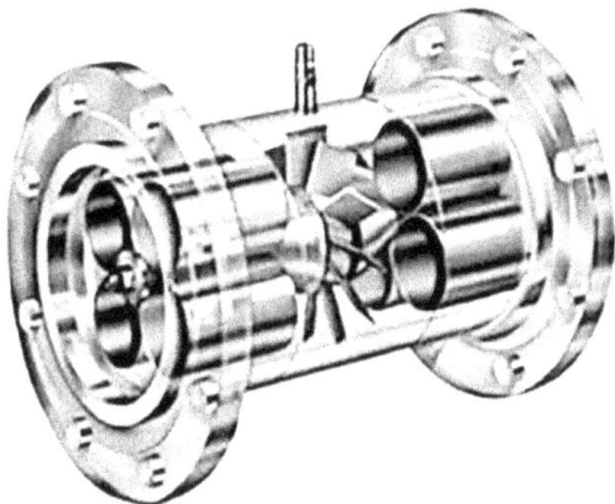

This type of meter is very linear, repeatable, can be made in steel, suitable for liquids and gases but not for vapor.

It can work at high temperatures and pressures and can be bidirectional.

4.3.3 Ultrasonic flow meter

The ultrasonic flow meter is a volumetric type of meter for liquids and gases. This measurement principle is the only technology that can measure a flow without being intrusive or requiring mechanical work on pipes. It has no moving parts and therefore has no maintenance, no load losses and is one of the few solutions capable of measuring bidirectional fluids.

Its operation is based on the propagation of ultrasonic pulses that travel through the liquid at a velocity specific to each material. Outside the pipeline two transducers are applied at a predetermined distance which, alternatively, emit a sequence of ultrasonic pulses, called a pulse train, and receive the train produced by the twin transducer. The time it takes for ultrasonic to go from the first transducer to the second and vice versa, is affected by the direction of the flow and involves two times of travel, the difference being proportional to the speed of the flow. Knowing the passing area, the volume is determined.

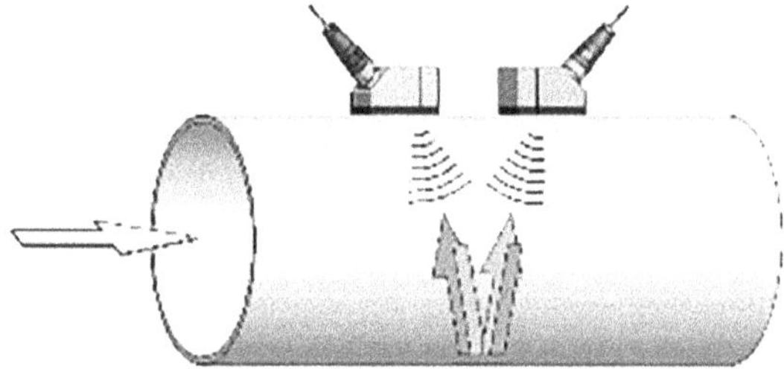

With the addition of two temperature sensors the ultrasonic flow meter transforms into a calorie count or fringe counts to calculate, for example, the thermal power exchanged.

The ultrasonic meter is composed of at least two sensors. At a low cost, it can measure most fluids.

It can be used with substances that do not need to be contaminated, since the sensors are located outside the duct and do not come into contact with the fluid to be measured.

4.3.4   Pressure differential flow meter

Flow measurement can also be achieved by measuring the pressure differential, with devices using the venturimetric principle.

In practice, the device has two chambers that fill with liquid or gas during operation. In the presence of a flow, the difference in their pressure acts on a membrane, allowing flow measurement. The two chambers are connected by two conduits that have calibrated bottlenecks and are placed in direct contact with the pipeline containing the fluid to be measured.

This type of flow meter can withstand high pressures and high temperatures, has a low load loss and is one of the most popular devices for measuring in plants, even outside. It can be a fully mechanical device, with a hand-operated indicator, or in a device with an electronic control device for interfacing to the electrical system.

Another system that exploits the pressure differential is to insert a calibrated diaphragm inside the pipeline and detect the pressure upstream and downstream of the diaphragm itself by two pressure transducers.

Below are stylized pressure transducers that detect the pressure difference existing at the two sides of the "d" diaphragm, which is used to narrow the "D" duct.

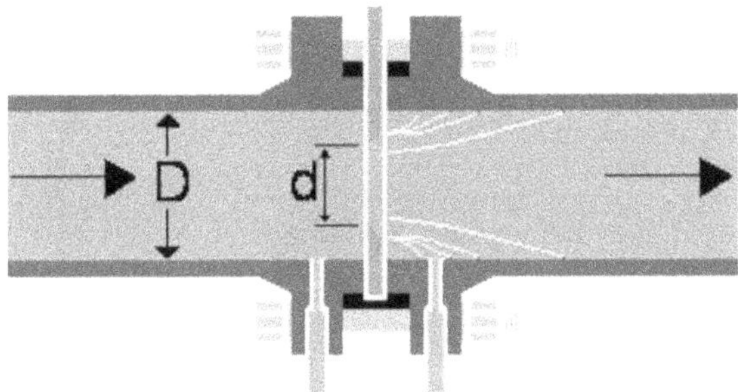

4.4 ACOUSTIC TRANSDUCERS

Rarely used in industry and in the automation of production processes, acoustic transducers are certainly the most popular type of transducer in the world.

Different types of acoustic transducers are **reversible**, i.e., they can generate an electric current at the detection of a change in pressure or generate an acoustic pressure wave if prompted by electric current.

These are transducers that we all use every day because they're installed in things like radios, televisions, telephones, alarm systems, vending machines. Their technology belongs to the past, the birth of these transducers takes place in the second half of the nineteenth century, but their use is more than current.

We're talking about **speakers** and **microphones**.

4.4.1 Carbon Microphone

It is the first type of microphone to be made. It uses coal dust to modulate the supply voltage, has high sensitivity but is not very stable and is affected by shocks and vibrations.

Still in use in traditional phones, home fixed line phones, it is practically no longer produced, replaced by modern electret microphones.

4.4.2 Magneto dynamic Microphone

Magneto dynamic microphones are the classic microphones used for audio shooting, live concerts, and studio recording.

They consist of a permanent magnet suitably shaped, in the magnetic field generated by the magnet is embedded a coil connected to a membrane. The vibrations of the membrane cause displacements of the coil, inducing a voltage proportional to the velocity of the coil.

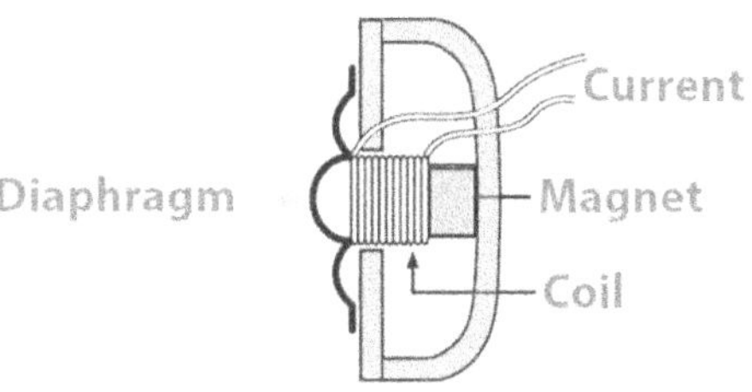

If mechanics are fine, these microphones have very little distortion. The response is affected by signal frequency, so equalization circuits are typically used to linearize the frequency response.

4.4.3   Piezoelectric Microphone

This type of microphone exploits the piezoelectric principle, so the voltage is proportional to the amount of deformation of the crystal and not to the deformation rate, so the vacuum output voltage is frequency independent.

These microphones are lightweight, less sensitive to electromagnetic disturbances and cheaper than dynamic microphones, but of a lower quality.

4.4.4   Condenser Microphone

It's the simplest microphone as a working principle. a metal box is closed by a metallic plastic membrane. The capacity that forms between the membrane and container is charged with a high voltage, while the vibrations cause the capacity to vary.

These microphones have a very high quality and are often used as sample microphones in acoustic measurements, or in recording studios. The high voltage required for their operation and the high sensitivity, which causes feedback and feedback effects, makes it uncomfortable to use live concerts, for which the use of magneto dynamic microphones is preferred.

The above pictures shows a condenser microphone for use in recording studios.

4.4.5   Electret Microphone

An electret microphone is a type of condenser microphone. Condenser microphones require electrical polarization. In the electret microphone, the polarizing charge is not provided from the outside, but is provided internally in the diaphragm or fixed capsule. Therefore, the microphone does not require an external power supply for bias.

An electret is a dielectric material with a permanent electric charge. It is created by melting an appropriate dielectric material containing polar molecules, and then causing it to re-solidify in a powerful uniform electrostatic field. The molecules of the dielectric align to the field direction, producing a permanent electrostatic "polarization". Modern electret microphones use PTFE plastic, polytetrafluoroethylene.

A pre-amplifier stage is required due to the low level of the microphone signal. Pre-amplifier stage power can be achieved either via a battery contained in the microphone itself or via the output cable, phantom power is in this case.

These microphones are very popular and cheap, the typical conformation is capsule: a small metal cylinder with built-in lead or soles for soldering conductors. the figure shows a typical electret microphone with wire already soldered:

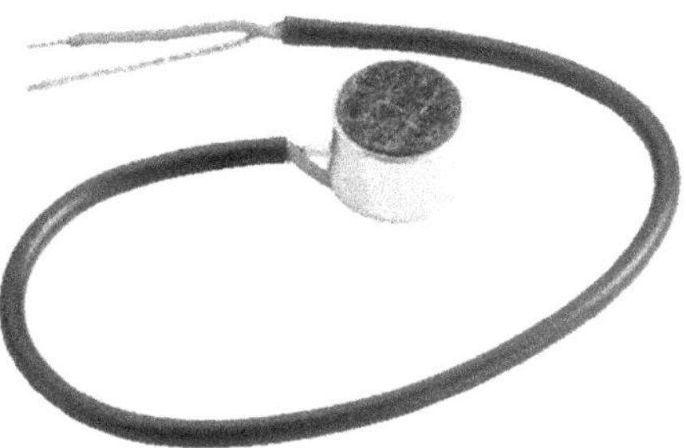

4.4.6 Magneto dynamic speaker

It works like a magneto dynamic microphone, so obviously the operation is reversed. you provide an electric current to generate sound, or sound pressure.

The sound is spread by a cone of paper or other material that moves together with the coil and is fixed along the perimeter, at the edge of the speaker, by a shaped rubber that allows the cone to move freely back and forth. The cone is used for low and medium frequencies and can reproduce relatively high frequencies, but for this purpose dome configurations, silk configurations, or special materials are more indicated that combine rigidity and lightness. The following figure shows a typical section speaker:

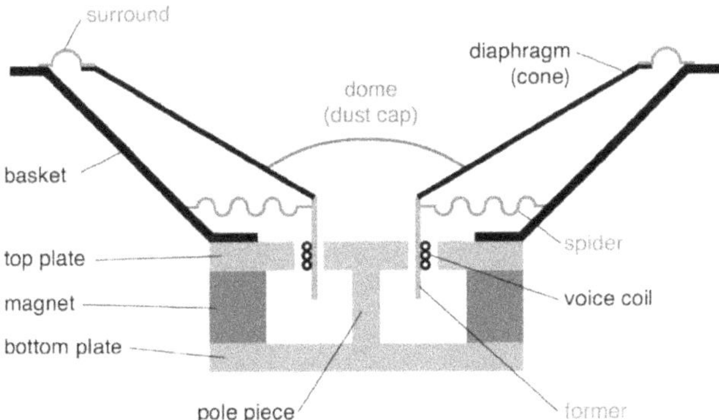

These loudspeakers arc used in all areas of audio, from live concert installations to small bookcase audiences. The characteristic feature of these speakers is that they have the maximum performance at their resonance frequency.

In order to reproduce the entire spectrum of frequencies audible to the human ear and to have a good quality at the same time, two or more speakers are usually used in the same speaker, in order to divide the frequency bands emitted by each of them. A low frequency specific speaker is called a woofer and a tweeter for high frequencies.

The magneto dynamic speaker, although high quality models exist, presents problems related to its electromechanical nature, such as distortion and phase variations.

The following figure shows a cone woofer speaker for hi-fi applications, with cardboard cone membrane and rubber suspension.

The next picture shows a tweeter speaker in a domed configuration for hi-fi applications, with silk dome:

The subdivision of the frequencies to be played is carried out by a circuit called a crossover, typically a simple passive circuit consisting of low pass and high pass filters, made by resistors, inductances, and capacitors. The crossover is placed in the speaker cabinet in which the speakers are housed, interposes between the amplifier and the speakers themselves.

4.4.7 Electrostatic diffuser

The speakers used for the construction of electrostatic audio speakers use a thin flat diaphragm, usually a plastic sheet coated with a conductive material such as graphite inserted between two electrically conductive grids.

By means of the conductive coating and an external high voltage supply the diaphragm is maintained at a potential of several kilovolts with respect to the grids. The grids are driven by the audio signal; the front and rear grilles are counterphase operated.

Consequently, a uniform electrostatic field proportional to the audio signal is produced between the two grids. This causes a force to be exerted on the charged diaphragm and its resulting movement pushes air on both sides of it. In almost all electrostatic speakers the diaphragm is driven by two grids, to minimize harmonic distortion.

The electrostatic construction is in practice a capacitor and the current is only needed to charge the capacity created by the diaphragm and the plates of the stator. This type of speaker is therefore a high impedance device. In contrast, a magneto dynamic speaker is a low-impedance device with higher current requirements.

4.4.8   Isodynamic diffuser

In the isodynamic diffuser the electric current, instead of going through a coil, runs a glued wire spread over a large sheet of material insensitive to thermal and electrically neutral excursions, usually mylar. In the absence of a signal, the mylar sheet is statically balanced in the magnetic field created by appropriately arranged magnets. When the electric current passes, the entire panel vibrates in a perfectly linear and thus free of distortion.

The isodynamic panel reproduces sound very faithfully and is used in acoustic speakers for high-fidelity audio systems.

In addition to the acoustic qualities, the isodynamic panel represents a resistive load close to the ideal for the amplifier, a resistance with negligible variability within the declared value. As a result, an extraordinary linearity is achieved across the entire reproduced acoustic range.

4.5 Humidity transducers

Moisture is always present in the air, in the form of water vapor. In measurements, moisture may be indicated as **relative** or **absolute**.

Humidity sensors can operate with the capacitive principle or the resistive principle.

4.5.1 Capacitive humidity transducers

This type of sensor, designed to measure relative humidity, is used especially in applications where cost, bulk and robustness are mandatory.

From a constructional point of view, a capacitive moisture sensor consists of a capacitor, among which is inserted an appropriate dielectric material whose dielectric constant varies with moisture. The most commonly used method for the fabrication of this sensor is a hygroscopic polymer film (used as a dielectric), at the ends of which the armor electrodes are deposited. The following figure illustrates a typical constructive example:

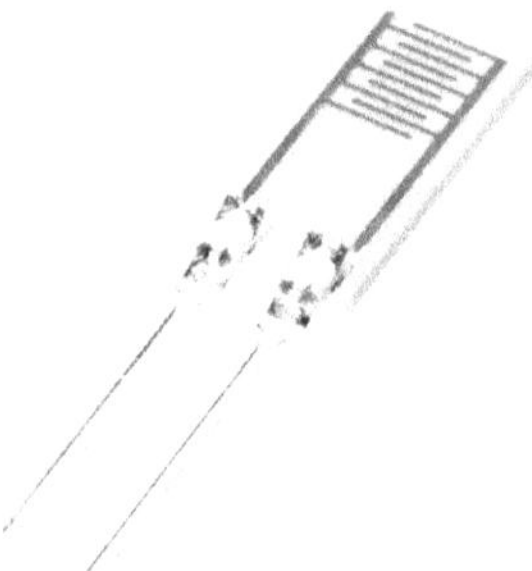

The advantages of capacitive humidity sensors are numerous; these include linear output voltage, high operating stability over time, wide range of measurement.

In order to obtain precise measurements of relative humidity, it is very important that the sensor and the material to be measured (gas) are at the same temperature. A moisture sensor cannot be encapsulated, as the measurement principle is based on the exchange of water vapor between the sensor and the gas mixture, e.g., air. The capacitive humidity sensor shall be positioned near the electronic control circuit.

The applications of capacitive humidity sensors are numerous: heating and air-conditioning systems, printers and faxes, weather stations, automotive, food industry, freezers, ovens and dryers.

4.5.2 Resistive moisture transducers

The principle behind these sensors is that conductivity, in non-metallic conductive materials, depends on moisture.

A resistive moisture sensor is usually made using materials with relatively low resistivity, so as to appreciate any variation in the moisture content according to the variation in the present moisture content. The relationship between resistance and humidity is inversely proportional.

The low resistivity material is deposited over two electrodes, which are arranged in a comb pattern in such a way as to increase the contact area as much as possible. The resistance between the electrodes changes when the upper layer absorbs moisture; this variation can then be measured and converted into a numerical value using an electronic control circuit.

Modern resistant moisture sensors are protected externally by a ceramic layer providing additional protection. Sensor electrodes are made of noble metals such as gold, silver or platinum.

The main advantages offered by resistive humidity sensors include low cost, compact size, high interchangeability as sensors do not require calibration, the possibility to place sensors at high distances to the electrical control and measurement circuit. The limit of these sensors is their sensitivity to chemical vapors and other contaminants. These sensors are used in industrial, domestic, commercial and environmental applications.

Absolute Humidity

It's the amount of water vapor contained in a volume of a mixture of gases. It can be expressed in kg steam per kg gas mixture.

Absolute humidity, depending on temperature, can rise to a limit value. Beyond this value water is separated into liquid form, the pressure at which water becomes liquid is called saturation pressure and depends greatly on temperature. The measurement of the absolute humidity of the air is not usually used because it does not consider temperature.

Relative Humidity

The relative humidity is obtained by considering the pressure of the water vapor existing in relation to the saturation pressure, expressed in percentage value (%RH). 100% relative humidity is the maximum amount of steam a mixture of gases can contain. The saturation pressure and, consequently, the relative humidity depend strongly on temperature.

Relative humidity decreases with increasing temperature. The above is valid for temperatures up to 0°C, below zero the saturation pressure over water and over ice should be distinguished, and the sensor documentation should be consulted for measurements at temperatures below 0°C. You need to know some definitions:

- The **dew point** is the temperature at which the gas must be cooled to condense the vapor contained in liquid form. It can be less than 0°C (32°F)
- The **freezing point** is the temperature at which the gas must be cooled to condense the vapor contained in solid form (ice). Frost point is a different temperature than the freezing point of the gas

Moisture in solids

Solids can also contain water and are called hygroscopic materials. The moisture content of a solid may be indicated as a percentage by weight.

If the vapor pressure contained in the solid is equivalent to the vapor pressure present in the air, there is an equilibrium situation. Any change in pressure results in an exchange of water until equilibrium pressure is reached between the air and the solid.

4.6 Gas transducers

There are several types of gas transducers, for different applications and with multiple principles of operation. We see some of the most popular models.

4.6.1 Semiconductor

The **semiconductor** gas transducer consists of a metal oxide. The detection of gas concentration is based on the measurement of the change in electrical resistance of the sensor in the presence of the gas with respect to the nominal value. The metal oxide has a very high resistivity at ambient temperature so that the sensor is operated at a temperature of about 400°C by means of a heater.

When the metal oxide comes into contact with air oxygen, a potential barrier forms between the oxide crystals that prevents the conduction electrons from moving freely. In the presence of reducing gases, the potential barrier decreases allowing electrons more freedom of movement and thus the resistance of the sensor is reduced.

In the figure is a semiconductor air-control sensor, such a sensor can be specific to detect, for example, LPG or ammonia:

4.6.2 Electrochemical Cell Sensor

Electrochemical cells are suitable for detecting many gases, whether toxic or hazardous, such as carbon monoxide (CO), hydrogen sulfide (H2S), sulfur dioxide (SO2), hydrogen cyanide (HCN), oxides of nitrogen, oxygen, hydrogen and others. Their operation is based on oxide-reduction reactions occurring inside the cell. Typically, the output signal is a current proportional to the gas concentration.

4.6.3   Catalytic sensors

They consist of a platinum filament coated with a catalytic compound formed by an inert base, e.g., alumina, and a catalytic metal, which accelerates the oxidation reaction. The filament is maintained at a temperature between 400°C and 500°C by a heater.

When a fuel gas is present in the air, it reacts with the catalyst to a controlled combustion, which increases the temperature of the filament; since this is platinum, the change in its resistance is directly proportional to the increase in temperature and the concentration of combustible gas is measured.

4.6.4   Infrared Sensors

Many gases are characterized by an infrared absorption spectrum with very narrow and non-overlapping bands, characteristic of the chemical element that constitutes the gas itself. This feature is the basis of the operation of the technology used to make infrared gas detectors, as shown in the figure.

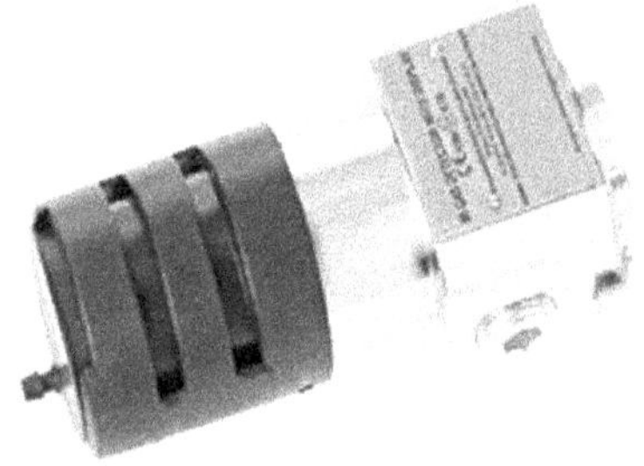

4.6.5   LASER spectroscopy sensors

There are some transducers on the market which detect the concentration of gas, such as oxygen, using the principle of LASER spectroscopy.

Spectroscopy is an analysis of materials based on the phenomenon of diffusion of a monochromatic electromagnetic radiation by the test substance.

These sensors can be used in a variety of applications, such as measuring the oxygen concentration in a gas, to make measurements in areas with explosion risk, for the verification of room ventilation.

4.7 IONIZING RADIATION detectors

About ionizing radiation

Electromagnetic radiation of a frequency exceeding 10^{16} [Hz] may be ionizing radiation if their energy, expressed in electronVolt [eV], is sufficiently high. An energy of 10-12 [eV] is considered sufficient to produce ionization.

Examples of ionizing radiation include X-rays and gamma rays. X-rays are commonly used in laboratories, medicine, and airport scanners. Gamma rays are used, for example, in the sterilization of medical components.

Ionizing radiation may have an effect on people's health, specific rules must be observed when working with machines producing ionizing radiation. Workers must be subject to regular health surveillance and their activity monitored by means of a "dosimeter" which records the amount of radiation absorbed over a period of time. Radiation can cause cancer, sterility and other serious illnesses.

Ionizing radiation does not make people or objects radioactive. Once the emission of radiation has ceased the environment is safe, radiation does not persist in space.

4.7.1 Ionization chambers

The ionization chamber is a capacitor that uses a gas as a dielectric. The capacitor is charged with a voltage of several tens of volts and in the absence of ionizing radiation is charged. When a charged particle passes through the gas, the ions it produces circulate towards the electrodes and so produce an output current.

4.7.2   Geiger-Muller Tubes

Geiger tubes are very high voltage operating chambers, up to 3000 [V]. The voltage is such as to cause an avalanche ionization effect to which is added the photoelectric emission of the electrodes, this results in the formation of a dense spatial charge of positive ions.

The tube usually contains low pressure gas.

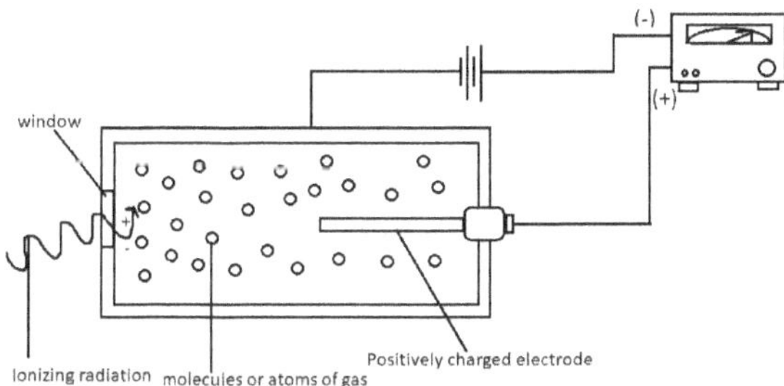

Geiger tubes do not measure the energy of particles but their number, have the advantage of having a very high output signal, of some volts. The instrument using the Geiger-Muller tube is commonly referred to as the Geiger counter and is used to measure radiation from alpha, beta, and gamma-type decays such as helium nuclei, electrons, and high-energy photons.

4.7.3   Semiconductor detectors

This type of transducer is essentially a photodiode. Ionizing radiation interacts in the emptying zone of the inversely polarized diode, the gaps and electrons produced create an electric current.

4.7.4   Scintillators

Scintillators convert the energy of X-rays and gamma rays into light.

Scintillators work with the principle of fluorescence, i.e., they emit light for a short time after being excited by ionizing radiation.

4.7.5   Brilliance Amplifiers

The luminance amplifier or brightness intensifier makes it possible to transform an optical image into an electronic image, is used in the medical world to increase the brightness and precision of a radioscopic image, thus allowing a reduction of the X-ray dosage for the same result. It is a useful tool to guide different medical and diagnostic procedures.

In the 1980s, new generations of brilliance amplifiers and cameras have enabled the production of digitized radioscopic images.

The brilliance intensifier is an alternative to classic photographic film.

The amplifier is an electronic tube between two screens and powered by electric voltage. The larger fluorescent input screen receives the poorly illuminated X-ray image, transforming it into an electron flow inside the tube. It is commonly called "IB".

The tension applied to the tube accelerates the electrons, which invest the second screen with an additional energy. The small output screen transforms the flow of electrons into visible light, yielding a much brighter image. The radioscopic image is then retransmitted on a monitor.

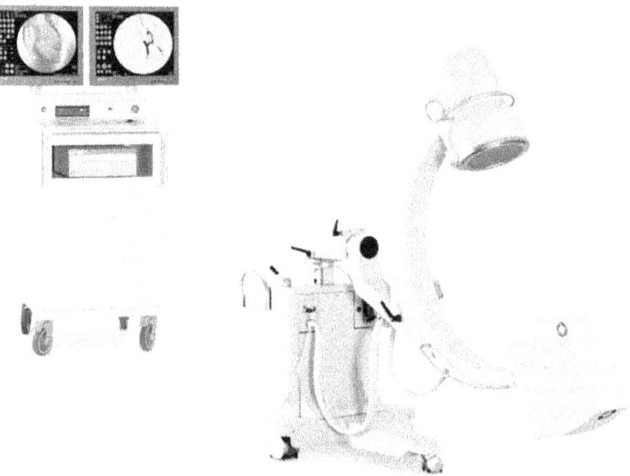

A portable fluoroscopy device, or "C bow", is visible in the photo and is commonly used in hospitals. The brilliance intensifier is the approximately cylindrical component visible at the top of the arch. At the opposite end is the parallelepiped monobloc that houses the radiogenic tube. The detected and intensified image can be viewed in real time on one or more monitors, generally connected to a PC.

4.8   POTENTIOMETERS

Potentiometer is a variable resistor normally used in electronics in the form of a trimmer, for use on electronic cards, or a potentiometer to be mounted on a panel for manual operation, by knob or linear cursor.

A potentiometer is composed of a resistive layer, which can be made of various materials, and a conductive material cursor that flows in contact with it. The electric conductors can be connected to the ends of the resistance and of the central slider.

Aspect of a classic graphite resistance rotary potentiometer:

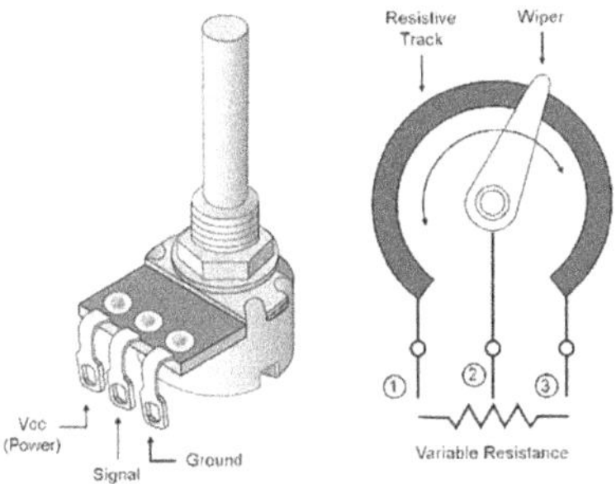

The electrical resistance between one end and the cursor varies as the latter moves. Measuring this strength provides the position of the cursor.

Potentiometers can be linear, such as audio mixer sliders, or rotary, such as a traditional radio unit volume knob. Their output can be linear or logarithmic. The linear version is for general use, the logarithmic version is used, for example, in audio applications and musical instruments.

An industrial linear potentiometer is shown below:

4.9   TEMPERATURE Transducers

There are many models of temperature sensors and transducers, with different operating ranges and different construction types. But almost all sensors are based on two fundamental principles: thermoelectric and thermoresistive.

Thermoelectric principle

The thermoelectric effect is the change in the voltage of a **metal junction** as the temperature changes. The device that exploits this principle is called a **thermocouple**.

This effect occurs at the junction between two different appropriate metals and is also called the **Seebeck effect**.

In practice, heating the metal junction, also called the hot junction, generates a difference in potential at the end of the conductors, cold junction, connected to the two metals of the junction. The measurable voltage depends on the material of the two wires and on the temperature difference between the hot joint and the cold joint.

The use of different metals in the junction allows to obtain different thermocouple models with different temperature ranges, output voltages and precision.

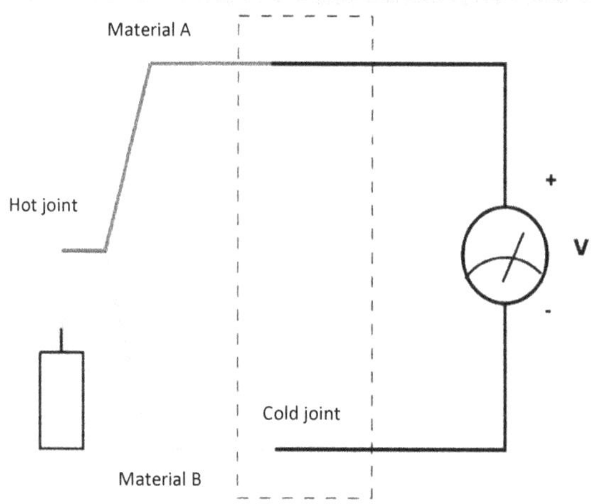

Thermoresistive principle

The thermoresistive effect occurs as the electrical **resistance** of a device, metal or semiconductor, **changes with temperature**.

The device that exploits this principle is called **thermoresistance** or RTD. The most commonly used materials are platinum and nickel.

Peltier effect

The Peltier effect arises when one circulates an electric current through the junctions of two different conductors, which absorb or produce heat depending on the direction of the current. The reversible effect does not depend on the nature of the contact but is related to the current in circulation. If the latter circulates in the same direction as the Seebeck current, the cold junction frees heat while the hot one absorbs it.

This effect is based on the **Peltier cells**, which are used for heating or cooling, depending on the direction of current flowing through them, the objects that are in contact with them.

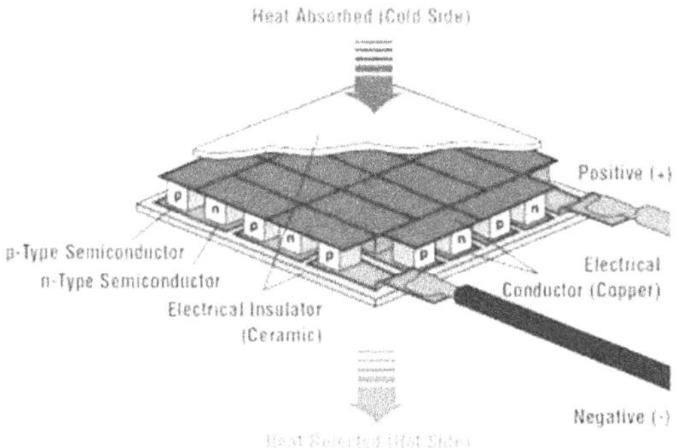

Cells are an example of a transducer converting electrical energy into another form: Heat.

Peltier cells are used to rapidly cool or heat small amounts of material. For example, to freeze biological samples, to cool CCD sensors, telescopes, and cameras, in LASERs to maintain stable working temperature, and in some contexts to cool computer microprocessors. They can also be used as a refrigeration system in small portable car and camper refrigerators, and in mini cold-water dispensers.

4.9.1 Thermocouples and thermoresistance

Thermocouples operate using the thermoelectric principle described above. They cover a very wide measurable temperature range, from about -200°C to about +1450°C. The various types of thermocouples are identified by a letter. Each model has a defined temperature range and a maximum output voltage, typically a few tens of mV, as shown in the table below:

	TYPE OF THERMOCOUPLE				
	J	**K**	**E**	**T**	**R**
T min [°C]	0	-200	-200	-200	0
Max. T [°C]	750	1250	900	350	1450
Max V [mV]	42	51	69	18	17

Thermocouples are inexpensive, are suitable for operation in heavy conditions, for example in very dirty and humid environments, and have the ability to measure high temperatures, in addition to being self-exciting, or produce the voltage alone. In contrast, they have very low output and are not very linear.

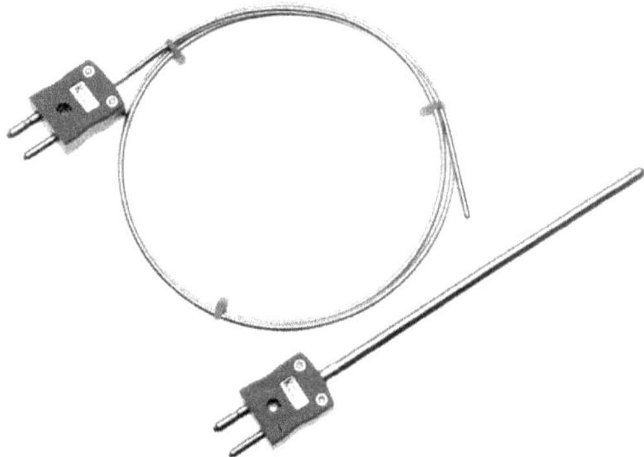

Two K-type thermocouples are shown, one in a flexible version, both with the typical two-pole connecting connector.

Thermoresistance are relatively inexpensive, very linear, fast in response, and also suitable for operation in hostile environments. They are more accurate and reliable than thermocouples. On the other hand, they require a feeding circuit and do not reach the maximum temperatures detectable with thermocouples.

A standard thermoresistance is PT100. It has a resistance of 100 ohms at 0°C, and a temperature coefficient of around 0.004. It can be used from approximately -200°C to +850°C. Example of PT100 in steel container with built-in cable:

Thermoresistance measure by CONTACT, must adhere, or be housed inside, to the component of which temperature is to be measured. To facilitate contact of all useful surface, silicone fats are often used between the body of thermoresistance and its location in the detail to be measured.

Temperature control systems

In many industrial processes, it is important that machining is carried out at a constant temperature, and it is often necessary to maintain a mechanical piece, or a demarcated area, at a defined temperature. To obtain accurate thermoregulation, a thermoregulator is a device called a thermoregulator, which reads the thermocouple signal and controls heating elements, usually resistors, accordingly. Modern digital thermoregulators can be very different from each other, in the figure you see a modern thermoregulator with buttons for setting operating parameters.

Temperature adjustments can be achieved by a variety of techniques, from the traditional ON-OFF method, which functions in practice like the old home thermostat, to more sophisticated control techniques with automatic learning of process characteristics.

Next, we see the different types of operation.

ON-OFF Control Operation

In heating operation mode, the heater enabling output shall be activated if the measured temperature is below the set value and deactivated when the threshold is exceeded.

This is the least accurate method, as the heater is always on at 0 or 100% of its power, removing or providing the voltage, without modulation. This method is also known as two-point control.

P Operation

In **Proportional** operation the output is not ON-OFF but proportional to the temperature input to the thermoregulator. The output can be at a value of 100%, if the temperature is very different from the set value and outside the range defined as proportional. As the measured temperature approaches the set value, the output is reduced until it is completely shut down.

Setting a very narrow proportional band can produce on-OFF-like operation, while a proportional band of adequate amplitude produces a much more stable controlled temperature.

I Operation

The **Integral** action is used to have an output value proportional to the integral value of the input time.

Setting the integration time correctly gives an output that is more or less stable. in particular with short integral time the output fluctuates rapidly around the set value before stabilizing, while with an integral time along the controlled temperature it comes close to the most softly set value.

D Operation

Operation in **Derivative** mode is used to have an output value proportional to the derived input time value.

Both the proportional and the integral action respond slowly to the change in temperature. The derivative action corrects the output value of the thermoregulator in a manner proportional to the inclination of the temperature change, quickly bringing the temperature under control to the set value.

Setting a long derivative time will oscillate the output value around the set value. with a short derivative time the system response will be faster, therefore, the most constant temperature value.

PID control is a combination of the Proportional, Integral and Derivative actions and allows for uniform temperature control, without large oscillations due to the proportional action, the adjustment of the offset obtained with the integral control and a rapid response to changes induced by external factors by the derivative action. For example, an external factor can be the opening of the door of a cooking oven, which causes a rapid drop in temperature.

A PID control can have many operational modes and settings. hysteresis, offset, proportional band, derivative time, integral time.

To accommodate users, many thermoregulator tools are equipped with **auto-tuning** or self-learning functions. The auto-learning function can automatically define the optimal parameters for the specific process. Often it is necessary and sufficient to leave the thermoregulator active, which automatically runs heating and cooling cycles, to automatically obtain the optimal values of the various parameters.

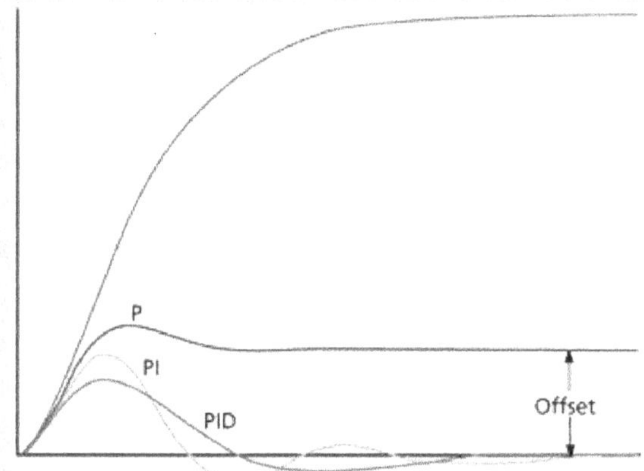

The graph compares the responses of a controlled process to the different methods shown. On the X-axis, time is represented, on the Y-axis, the value of the variable to be controlled, such as temperature.

The highest curve represents the process without control. It is clear that PID regulation can contain the oscillations present in PI control and that the offset is practically zero compared to simple P-type regulation.

4.10 MAGNETIC POSITION TRANSDUCERS

The devices made using the magnetic operating principle are used for **position, distance, speed and acceleration detection**.

4.10.1 MAGNETO STRICTIVE TRANSDUCERS

Transducers using the principle of magneto strictive operation generally consist of a cylindrical rod and a hollow cursor flowing along the rod. The electrical part shall be placed at the end of the auction. Inside the rod itself is a special conductive cable called a **waveguide**, while inside the cursor are one or more magnets.

A current pulse is sent within the waveguide to generate a corresponding magnetic pulse. This magnetic field interacts with the field of the magnets contained in the cursor and generates a mechanical deformation pulse. This impulse propagates at the speed of sound, which in metals is about 2800 [m/s], until it reaches the sensor in the head containing the electronics.

The measurement of the **time** elapsed between the initial electric pulse and the mechanical pulse of return provides the position of the cursor, i.e., the area where the two magnetic fields have overlapped.

The construction materials must be non-magnetic, usually steel or aluminum. External magnetic fields generated by other magnets, or the proximity of the cursor to magnetizable materials, can impair measurement.

4.10.2 MAGNETO RESISTIVE TRANSDUCERS

Transducers using the magneto resistive operating principle are ferromagnetic material (Ni-Fe-Co) magnetized in a predetermined direction such that the action of an external magnetic field causes a rotation of the initial magnetization and therefore a change in the resistivity of the conductor.

4.10.3 LVDT transducers

Linear Variable Differential Transformer. Transducers using this principle are made up of a hollow cylinder within which a mobile part moves. The cylindrical container shall be enclosed with a primary and two secondary measuring windings. The mobile crew consists of a rod of metallic material that flows through the cylinder without contact. A power and control circuit is required.

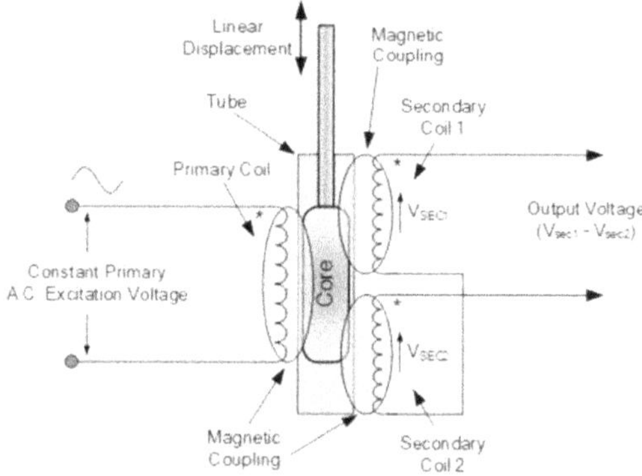

The primary winding of the LVDT is powered by an oscillator in alternating current (see figure). The magnetic flux produced is coupled, through the mobile crew, with the secondary windings. The difference of the induced voltages on the secondary windings (Vsec1 and Vsec2) is proportional to **the displacement of the mobile part** with respect to the central position, to which there is zero voltage.

The advantages of a LVDT type transducer are very long life (very low friction), high resolution, high sensitivity, absolute measure of motion, and zero position repeatability.

HALL effect

The Hall effect is the phenomenon of electric current induced by a simultaneous magnetic field. This can occur in metallic materials or semiconductors.

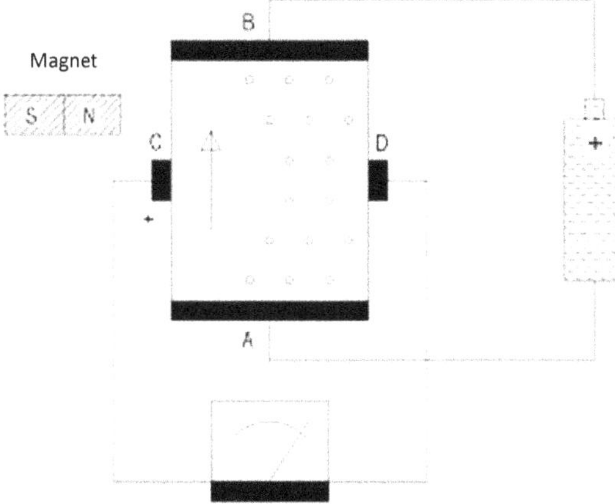

When a continuous voltage is applied to the leaders of a conductor, an electron flux (A-B) is produced. If a magnet approaches the conductor at this point, the electron flux undergoes a deviation from the straight path, with a thickening towards the point C or D, depending on the polarity of the magnet, resulting in a difference of undetectable C-D potential without the magnet.

With this principle of operation are realized both analog output transducers and ON-OFF sensors.

4.11 PIEZO transducers and extension meters

The term piezo comes from the Greek, meaning *to apply pressure*. Transducers using this operating principle are devices that exploit the properties that some materials show when pressed by mechanical action.

4.11.1 Piezoelectric principle

The piezoelectric effect is the property of some materials to **generate electric charge** when subjected to mechanical stress. These elements are electrically neutral, but if a force acts on them, their structure deforms to produce excess charge at the surface. This phenomenon is reversible so that the application of a potential difference will produce a mechanical deformation.

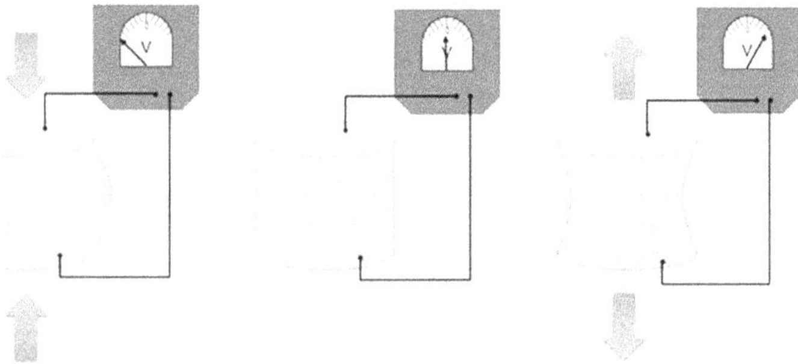

Note that the effect is not observed in the absence or in the presence of static stress, but only in the case of **changes in stress**.

Transducers that rely on the piezoelectric principle are materials such as quartz, lithium sulfate, tourmaline, polymers, or ceramic materials, which, having no center of symmetry in the crystal structure, exhibit electric dipoles. Mechanical stresses modify the dipole moments, resulting in the generation of electric charge differences.

The piezoelectric coefficients of different materials are temperature dependent.

Some examples of transducers built with this principle are sensors of force, pressure, vibration and acceleration (accelerometers).

4.11.2 Piezoresistive Principle

The piezoresistive effect describes the peculiarity of many materials to **vary** their **electrical resistance** when subjected to mechanical deformation. Mechanically strained, these materials keep their volume constant, but vary their section and length; as a result, their electrical resistance will vary. Semiconductor materials may be used for this purpose.

Examples of the use of transducers built with this principle are strain gages, accelerometers, microphones.

4.11.3 Extensimeters

The strain gage is a very thin **wire,** usually Constantine (a copper-nickel alloy), applied to a support of plastic material. The strand of the extensometer shall follow the deformations of the surface to which it is attached, stretching and shortening along with it; these changes cause a **change in the electrical resistance** of the wire and by measuring these changes it is possible to trace the extent of the deformation that caused them.

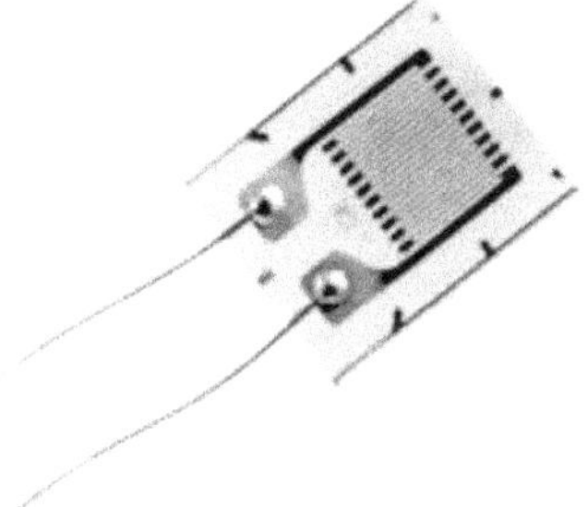

The electric extensometer, also called **strain gage**, is a highly sensitive transducer with a high response speed. One of its typical applications is the **load cell**, used in weighing systems.

The strain gage can also be made using semiconductor materials. In this case, the **piezoresistive** principle is used.

The classical configuration is that with Wheatstone bridge connection, in practice 4 elements are connected to form a square, to the two opposite vertices is applied the supply voltage, from the other two vertices is taken the output signal that gives the amount of deformation suffered by the sensor structure.

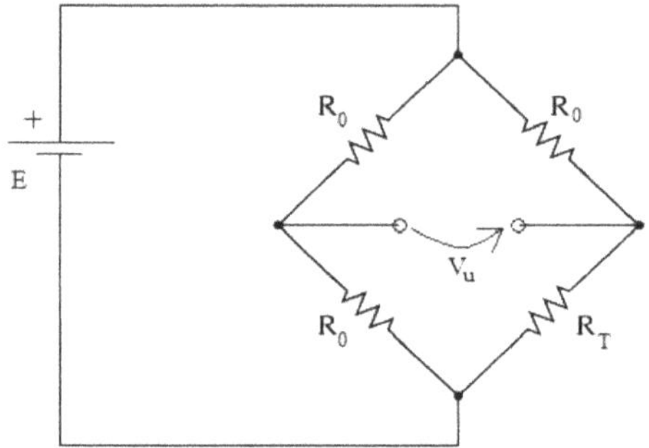

Piezoelectric sensors have several advantages compared to strain-gage sensors, namely: very wide range, sensitivity, reaction threshold and resolution independent of measurement range, compact design, stiffness and overload protection, long operating life, wide operating temperature range, low interference sensitivity.

4.11.4 Load cells

This classification includes transducers made with one or more extensimeters connected together in the same sensor.

Typically, the load cell requires a supply voltage and provides an output proportional to the value of the supply. This proportion is commonly called **sensitivity** and is expressed in **mV/V**.

An electronic board is usually needed to handle the cell signal and to make the various compensatory calibrations of 0 (offset) and span (gain). It is advisable to correct nonlinearities and compensate for variations due to temperature.

Example: A load cell is powered at 10 Vdc, its sensitivity is 2 mV/V, so its output will be 2 millivolts for each voltage supply, in this case 10. Its maximum full-scale output will be 20 mV if powered at 10 V.

There are load cells for traction and compression measurements, and bridge transducers can typically also be used for other measurements, e.g., **pressure sensors**. Below is an example of aluminum body load cells:

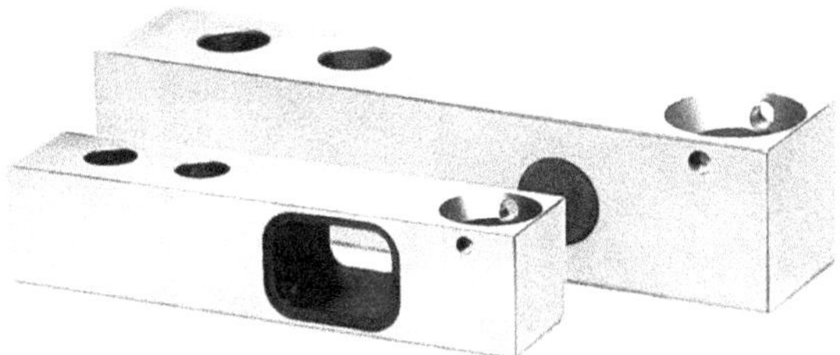

There are full-deck struts, consisting of four independent sensitive grids, oriented around the sides of a square. The four grids are connected to Wheatstone Bridge. These type of strain gages are particularly suitable for **torsion measurements**: for example, they allow to measure the torque transmitted by an axle and then, when coupled with a tachometer, to determine the power developed by an engine. This type of transducer is called a **torsiometer**.

5   OTHER DEVICES

In this chapter we find complex sensors that are not just about presence or distance detection but are part of more articulated systems, in which software is essential for correct use.

Some of these sensors rely on principles of operation already seen in this guide, such as smartphone sensors. Others, such as those for cameras and vision systems, share similar technologies and become specialized only with the contribution of the specific software.

The technological development in these devices is very fast and will probably be present on the market different sensors than those described here, developed over the time period that I have used to complete this guide.

5.1   Vision Systems

Vision systems are special equipment that allows the detection of various characteristics of an object. This paragraph provides only general guidance, a comprehensive treatment of the features of such systems would require a separate volume. It is appropriate to check directly with the manufacturer whether a vision system is being used and whether the most suitable model is chosen according to the desired detection. A good system can cost several thousand euros.

With a vision system it is possible to determine shape, color, measure, verify the orientation of an object, select between objects of different types or distinguish between pieces of the same type, a good piece and a waste. You can also perform identification operations such as reading bar codes or matrix codes, reading writings, or identifying a previously stored logo.

There are a number of interfacing possibilities, and these depend on the type of system used, the models are very different and each manufacturer has its own particular characteristics.

It is possible to communicate with a vision system using digital signals (PLC inputs and outputs), by connecting via a serial port or by using an Ethernet-type connection with the classic RJ45 connector.

Commercially, there are a variety of models, but typically one system consists of:
- digital camera, complete with a lens and any optical accessories. Some cameras can even be mounted on mobile systems.
- Illuminator, which can be either front, mounted on the side of the camera, or rear, to illuminate the part against the light. They are usually made with a high brightness selected LED matrix, are very expensive and sometimes require a dedicated power supply (12Vdc or more).
- control and interfacing electronics that can be housed in the body of the camera itself,
- monitor to view camera images and control parameters: numerical data, functions, areas of image to be analyzed. It can be integrated into the system or external, or even substituted with a PC, which can be used during tuning and detached during automatic operation.
- management software that can be internal to the electronics with integrated monitor or installed on a laptop. Once your vision system has been set up, it stores the settings in the system and your PC is no longer needed.
- cables and accessories: mounting brackets, bulkheads for protection from light, loudspeakers.

The following figure shows a vision system with built-in camera and front LED illuminator:

When choosing a system, you have to consider a few things:

- What do I have to see? Shape, color, measuring, reading codes or writing?
- Do I have to compare different pieces, or do I have to distinguish and sort all the same pieces?
- Are all the parts in the same position or in scattered order (e.g. on a conveyor belt)?
- How large is the area I need to examine and at what distance can I position the camera?
- Where can I place the illuminator?
- What processing speed is required? There are several controls that I can do, and each control takes a different processing time, and I can perform multiple checks on the same piece, which can increase the time taken
- What do I connect to my system? PC, PLC, monitor with composite input or Ethernet, data collection if applicable, etc.
- You need to always view this information, so do you need a dedicated monitor?

The answers to these questions can help identify the system best suited for the purpose.

An example of a vision system with two cameras connected to the control electronics:

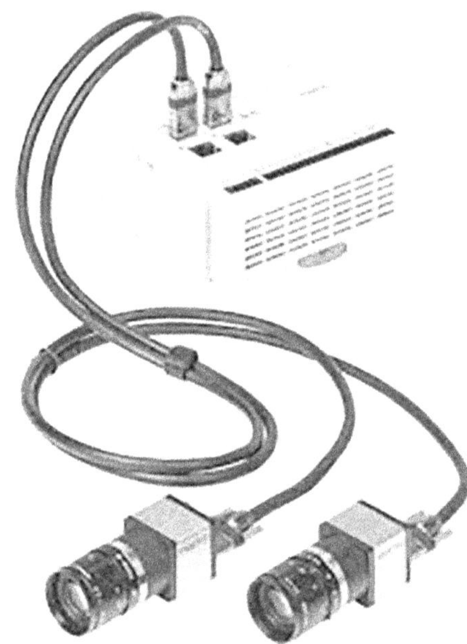

In practical use a vision system takes a picture of the part to be detected that is displayed on the monitor. The proprietary software then allows you to choose from a set of preset controls which you want to use. For example, shape control, border detection, measurements, color control, decode barcodes, or write reading. Usually, these operations are done in black and white mode, but color cameras and systems are available if required.

See example of a vision system. The camera detects the floppy disks on the fly, a PC processes the information and judges whether the component detected is good or bad, by size, shape, color, or due to defects or missing writing, incorrect positioning, or other. In case of discarding, a mechanical arm pushes the discard into a container:

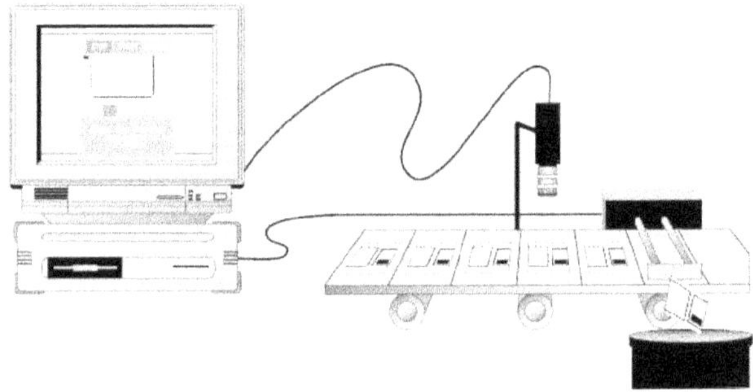

It is also possible to compare with sample images previously stored. After processing, depending on the outcome, the part may be managed by the installation: sorted, selected, discarded, or sent to later jobs.

A key thing in a vision system is the correct choice of lighting and its positioning.

In addition, it is a good practice to provide protection of the inspection zone from outside light, to avoid variations in camera readings. Potential disturbances may come from window light, or ambient lighting.

Provide for the possibility of turning off the illuminator, even in automatic mode, if it is not possible to do so directly from the vision system itself. This prolongs the life of expensive LEDs and avoids visual disturbance when maintenance is performed.

5.2   Barcode readers

Code readers are special sensors for reading different types of barcodes. There are also point matrix codes that can contain much more information for the same surface area used.

Code readers can be classified either portable or fixed. They can operate with LASER and CCD technology; LASER players can be single-scan or raster.

For example, fixed players are used in factories (production and distribution), supermarkets and video distribution units. Mobile players, also known as pistols, can mainly be used in warehouses or production departments.

LASER readers have a rotating polygonal mirror inside them. The LASER beam is sent to the rotating mirror, which produces a line on the barcode. Reflected light is read by a photodiode and converted to an A/D signal to highlight bright and dark areas. Decoding light and dark spaces, through the specific rules of each type of code, generates the code read. The code read is generally sent to the control electronics via a serial interface of type RS232 or other.

CCD-based readers illuminate the entire code with LED light, and the reflected image is then captured by the sensor as a single image, decoded, and transmitted in the same way as LASER readers.

For LASER readers the difference between single scanning or raster is the number of lines (rays) scanning the code. Raster scanning uses multiple lines, is therefore indicated for reading printed codes not optimally, even with partially incomplete or broken or stained lines, the important is that at least part of the code is readable and correct.

The choice between single scan or raster depends on the orientation of the labels with respect to the reading radius and on whether the codes to be detected are static or moving.

If you move the codes in the direction of "picket fence" a raster type reader is recommended. When moving in the "ladder" direction, a single-scan reader can be used, which then, with moving code, acts as a raster. Please refer to the "Code Structure" for explanation.

If the movement of the code may have to stop during the reading phase, it is recommended to use a raster type reader.

Code structure

A barcode is a series of parallel lines of variable width and spacing. There's a lot of barcodes, but basically all of it is based on the same structure.

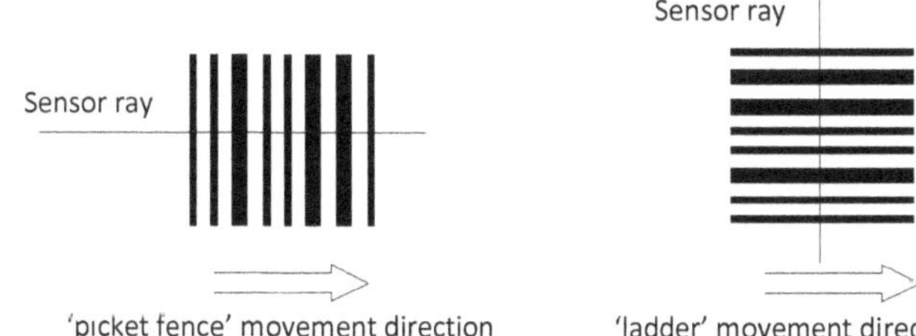

The structure provides a 'quiet zone' with no lines, at the beginning and end of the code, plus a series of lines used as Start Character, Data, Control Code and Stop Character. Depending on the encoding type, only numbers or the entire character set can be stored. The most widely used codes in the industrial field for product identification and marking are UPC/EAN/JAN and CODE39.

The codes can have very narrow and narrow bands, and the print quality is very important to get a correct reading. The best quality is obtained with LASER printers.

The demand to place a large amount of data in a barcode has led to its evolution, with high-density encodings such as CODE128. This evolution has led to a consequent upgrade of the players, which have become increasingly complex and performing.

In response to the need to store more data, the 2D two-dimensional code was developed, commonly called **QR code**.

The QR code used today, even in its Micro QR code, is derived from the first QR code released in 1982: the True Code and the later Data Matrix. Its structure is a matrix of points, or rather of squares. It is used with different types of encoding, array size, detection patterns, synchronization and error correction.

Compared to barcode, QR has:

- It also uses 1/30 of the space, which is a huge reduction in size.
- large amounts of information, over 7000 numbers and 4000 characters,
- readability in any position,

- automatic data recovery function in case of partially dirty or damaged code up to 30% of area.

This is an example of QR code with text, numbers, and special characters. Try to read it with your phone, there are several free apps that allow you to read QR codes using the camera built into your phone.

Did you manage to read it? Well, now try to cover a small part with a finger and read it again.

5.3　Camera sensors

Sensors used in cameras or mobile phones can be thought of as a grid of photo receivers, or light-sensitive elements. Regardless of the CMOS or CCD construction technology, what characterizes these sensors is the way individual photoreceivers, or photodiodes, are operated to achieve the digital image.

The receiver matrix, or grid, is usually arranged to form a rectangle. Individual receivers can be imagined as individual pixels in the image. The total number of pixels in the sensor is simply the base pixel product for the height pixels. Thus, for example, a sensor of 6000 base pixels and 4000 heights will produce an image of about 24 (Mega) million pixels, this is usually referred to as the resolution of the camera or phone.

The photodiodes give a signal proportional to the amount of light received, thus forming an image that very faithfully represents the variations in brightness but has no color information. In practice, the camera's sensor can only output black-and-white images.

Therefore, a mindset is required to achieve color images: it overlaps the photodiode grid of an optical filter made up of a grid formed of colored tiles, which obviously allow the passage of light, which overlaps perfectly with the grid formed by the photodiodes.

The figure below shows the photo diode grid, represented by a series of all equal tiles, with the colored matrix filter superimposed.

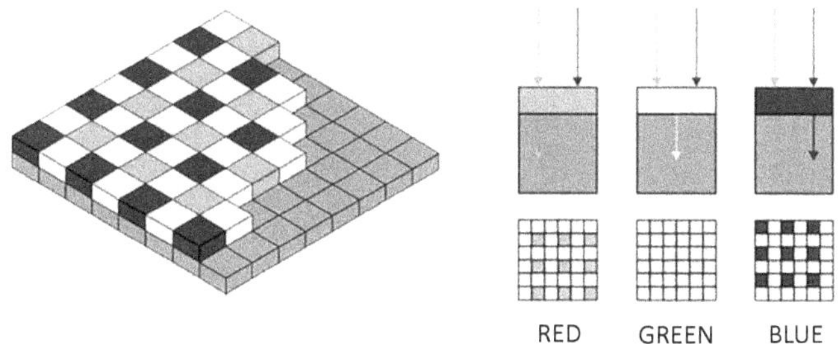

The various 'tiles' of this filter are usually three colors: **green** (the most numerous and lightest in the image, **red** and **blue** (the darkest). Green tiles are in twice as much quantity as red and blue ones, and they alternate; this type of distribution of colored filter tiles is called Bayer matrix or simply **Bayer filter**.

The application of the filter on the sensor transforms the image in black and white into an image with the three colors, green, red and blue with different intensities depending on the light that arrived at the sensor.

The filter itself, therefore, does not solve the problem, the image is still unusable and does not correspond to the scene photographed. In order to compose the scene respecting both the luminous intensity of the individual pixels and their color, an elaborate mathematical process is required which, considering the luminous intensity and color of the individual pixels and comparing it from time to time with the adjacent pixels, reconstructs by successive interpolations the 'real' colors of the scene.

This process is called **demosaicization** and allows to correct the raw image provided by the sensor in an image that respects, with good approximation, the colors and brightness of the scene photographed.

The Bayer filter is not the only one possible, in the past there have been numerous attempts to produce filters with different tile arrangements and using different colors. Big industry players have experimented with filters that include white, yellow and red tiles. However, each variation in the arrangement and coloring of the filter tiles requires a different demosaicization algorithm.

Today, most sensors use Bayer filter. The only exception in commercial products is the X-Trans matrix developed by Fujifilm. In this matrix, which uses the usual red, green and blue tiles, the tiles have a completely different arrangement.

Comparison between Bayer Filter and Fuji X-Trans Filter:

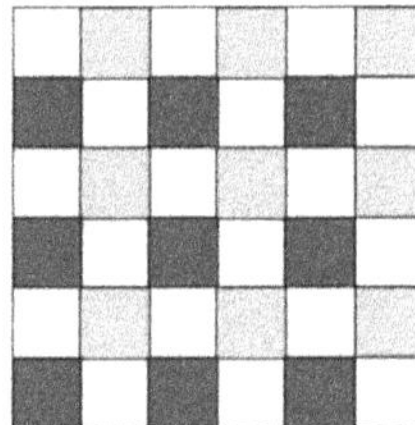

BAYER

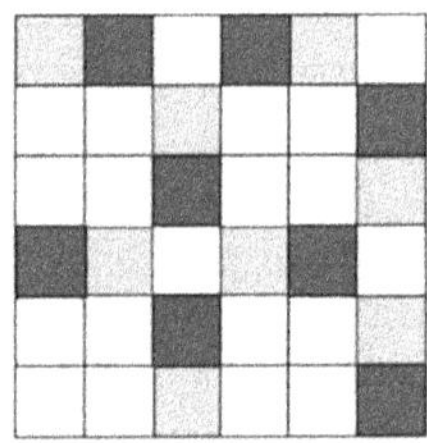

FUJI X-TRANS

The particular arrangement of X-Trans filter tiles, and the resulting use of a fully reworked demosaicization algorithm, brings some advantages over the Bayer matrix. The most obvious is the elimination, or at least the strong attenuation, of the so-called Moiré effect. This phenomenon involves the appearance of unreal colored areas when photographing subjects with repetitive wefts such as a fabric drawing of a jacket or sofa.

The sensors are manufactured in a variety of formats, regardless of the number of pixels and the filter used.

The figure below shows the size of the different sensors by comparison, the image is double the size of the real. The "full frame" sensor refers to a sensor that is exactly the size of an old slide, that is 24 x 36 mm on a side, usually used in professional and expensive cameras. Larger sensors are also available, but their use is limited, usually one inch (2.54 mm) diagonal sensor used in most compact cameras.

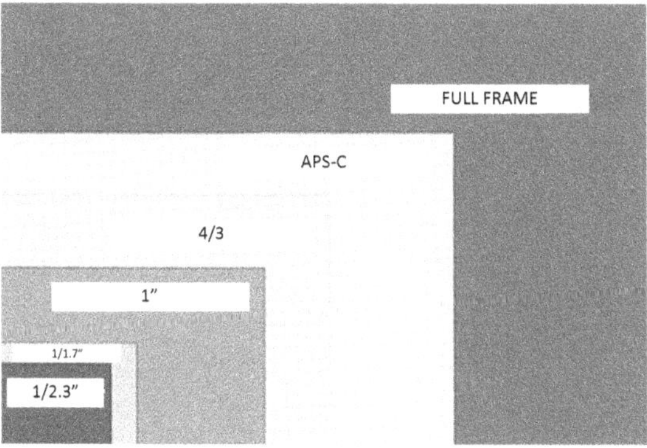

This is the appearance of a sensor ready to be mounted in a digital camera, complete with filter, protection glass and electrical connections:

5.4 Smartphone sensors

All the operating principles and the types of sensors seen so far are also usable in common and everyday contexts, I am referring to the use of modern smartphones.

Miniaturized systems, also known as MEMS Micro Electro Mechanical Systems, are used in 'mobile phones', as well as in video game consoles, which take advantage of modern technology to produce miniaturized objects.

These devices can be mechanical, electrical or electronic and are usually bundled together into a single integrated circuit that becomes, for example, a pressure sensor or an accelerometer as needed. Applied to the phone and paired with the apps that run them these systems become proximity and brightness sensors, gyroscope and accelerometer. Not all phones all have sensors described, it's very much dependent on price range.

accelerometers measure accelerations, typically in units of g, where $1g$ is the acceleration due to Earth's gravity and is equal to about 9.8 m/s^2.

5.4.1 Ambient light sensor

This sensor provides a current proportional to the amount of ambient light it detects.

It is used to automatically adjust the brightness of the phone's screen according to the amount of ambient light, making viewing more comfortable. The sensor can support automatic white balance (AWB) when taking photos and also interacts with the proximity sensor to avoid usage errors when, for example, the phone is in the pocket.

5.4.2   Proximity Sensor

It consists of an infrared LED and a receiver and is generally located at the top of the screen. Used to detect the distance between an object and the phone. The operating range of a proximity sensor is generally ten centimeters.

Detects nearby objects without contact and automatically switches off the screen when it detects that it is close to the ear.

5.4.3   Gravity sensor or accelerometer

The accelerometer measures changes caused by movement on three axes (X, Y and Z) and determines acceleration and deceleration forces.

It allows the phone to automatically switch between horizontal and vertical modes, count daily steps, identify display orientation, use the compass (App) and recognize motion gestures (such as lifting and turning the phone). It can be used to silence an incoming call by flipping the phone.

Often used by Apps in conjunction with the gyroscope, in models that feature both sensors, this sensor has revolutionized the use of smartphones.

5.4.4   Gyroscope

A gyroscope is a device consisting of a wheel spinning in a fixed structure so as to lean freely in any direction. The angular momentum of the wheel allows it to retain attitude when the structure is tilted, thus allowing it to measure or maintain orientation and angular velocity. Typically, a triple-structure gyroscope is used on the phones, which can simultaneously determine position, motion tracking, and acceleration in six directions.

The gyroscope allows games to be used, moving the phone to change display, navigation using GPS, and virtual reality apps.

It can be used in conjunction with the accelerometer to run Apps like the Google Sky Map showing the constellations, stars and planets in the direction we point to the phone.

5.4.5   Compass

The compass consists of a magnetic field sensor that measures the strength and direction of a magnetic field by detecting changes in the electrical resistance of magneto resistive materials within it. You may need to shake the phone in order for the sensor to work properly. To ensure the accuracy of the measurement results, do not keep the phone close to magnetic objects.

The compass is used for map navigation, to help locate locations more accurately.

5.4.6   Hall effect sensor

In a Hall effect sensor, a current is applied along a conductor. When the conductor is immersed in a magnetic field perpendicular to the direction of the conductor, electrons will be deviated from the rectilinear trajectory they would have in the absence of electromagnetic fields. As a result, one side of the conductor will charge negatively and the other side will load positively, causing a potential difference.

This sensor can be used for "smart cover" functions that automatically shuts down the screen when the cover is closed, which has a small magnet inside it.

5.4.7   Barometer

Consisting of a rheostat and capacitor, a barometer measures atmospheric pressure by calculating changes in electrical resistance and in capacitance.

It is used to correct altitude measurement errors and interacts with the phone's GPS to locate the position when the GPS signal is weak or absent, as when located indoors.

5.4.8   Pedometer and heart rate monitor

They are the typical sensors used by those who play sports and want to monitor the activity carried out. The pedometer function is actually often performed by apps that use the accelerometer to simulate the pedometer and count steps.

The heart rate monitor is rarely seen on smartphones. A cardiofrequenzimeter is a device that is worn, in the form of a chest band or bracelet, and features electrodes to measure the heartbeat, data is transmitted to the phone usually via Bluetooth.

5.4.9   Thermometer

Like all devices, smartphones also produce heat. In certain situations, heat can reach dangerous levels for the mobile phone. The temperature must be constantly monitored to prevent the device from overheating and damaging. The thermometer, always present inside smartphones, in combination with the software of the phone itself, can serve to limit the performance in case of overheating.

5.4.10 Fingerprint Reader

First introduced only in high-end smartphones, it is now found in mid-range models as well. Used for accessing programs or websites, instead of a password, to ensure more security for users.

The fingerprint sensor performs fingerprint scans. Compared to a common password, the fingerprint reader provides a higher degree of protection and is quicker to use than an alphanumeric password.

5.5   Measuring systems

With modern technology it is possible to make optical measuring instruments that are extremely fast and accurate.

A LASER is usually used for beam or measurement beam generation, but a high intensity LED diode may also be used. A CCD or LASER receiver can be used for reception.

Measurement systems can detect the size of small objects by using the barrage principle. The emitter develops a beam of light, which must be perfectly aligned with the receiver. If an object interposes itself to this beam, it introduces a shadow area that is detected by the receiver.

Such systems can also be used for eccentricity detection, scattering, position, edge detection, continuous inspection of dimensions.

In the figure you can an optical micrometer for measuring, in this case the diameter of a pin, with its viewing and control tool. The LASER beam, usually not visible in the air, which is emitted by the emitter module to the left, is shown.

With systems of this type, it is possible to detect, with precision of the order of 1 [μm] or less, the dimensions of objects up to about 100 [mm]. The projector and the receiver can be positioned at a great distance, on the order of a few meters, without compromising the measurement characteristics of the system.

5.6   Quantum sensors

At the time of writing, January 2021, the development of industrial quantum optical sensors is being studied. Quantum technology will soon be available for industrial sensors. A functional test of the world's first quantum optical sensor for mass production has already been performed.

Quantum sensors would allow measurements with unprecedented precision. Using laser light, quantum sensors allow highly efficient measurements that would be impossible with conventional processes.

So far quantum sensors have been used primarily in research.

Quantum technology enables very fast measurements of particle movement and dimensional distribution. Industrialization of these sensors takes an important step towards commercialization of quantum technology.

Quantum technology shifts the technical limits that have been set. Using quantum effects, additional details can be perceived from signal noise where, until now, no specific signal would be measurable. This allows the measurement of particles that are about two hundred times smaller than the width of a human hair. Quantum sensors will initially be used to analyze substances in the air.

Quantum sensors could in the future become everyday equipment in various sectors: for example, they could be used in civil engineering to visualize underground structures before construction works commenced; in the electronics sector circuits can be inspected across surfaces; Extremely accurate measurements may be made in industry in general.

5.7 Domotic Sensors

Domotic is the automation and control of a building, by means of sensors and a management electronics. With a domotic system it is possible to control the ambient temperature, the level of brightness, any leakage of water or gas or to carry out an intrusion control and its alarm system or to keep control of the consumption of the users.

Home automation systems can be remotely connected to control or control the home remotely, via the Internet or a smartphone app. For example, the air conditioning or heating system can be turned on or off using a temperature sensor, or windows can be closed or opened with motor power according to the brightness level of a sensor. Thanks to the use of the home, it is possible to better manage openings outside the apartment (shutters, sunblinds, motorized doors/windows, skylights, etc.) or monitor the electricity consumption in real time and provide for the remote-controlled ignition or shutdown of specific users, for example a domestic appliance.

5.7.1 Power saving

With home automation, you can reduce energy waste. A domotic electrical load control device allows to manage the power of the system and automatically disconnect household appliances in case of overload, thus avoiding the annoying disguise of the blackout and reducing peak consumption.

An evolved device can allow for the differentiated management of the activation of loads according to a consensus from a clock, then there is an intelligent management of electrical loads in such a way as to promote energy savings by using the most appropriate time slots.

You can consistently view your electricity, water and gas consumption in real time, useful information to learn how to make the most of resources, reduce waste and evaluate any malfunctions.

Energy management sensors can be pulse emitters (flow meters) for connection along a water or gas line or a T.A. (current transformer) for connection along the electrical distribution.

In the following figure you can see schematized the connection of a TA, to detect the current leaving the counter and check some appliances in this way. This example shows a washing machine, a dishwasher, an oven and a water heater connected to power sockets controlled by a domestic device. The TA is connected to a detector which, via a bus (represented by the wires woven), gives the indication of consumption to the device controlling the wall sockets. The refrigerator, last to the right, is connected to an outlet not controlled by domotic, as it should not be turned off.

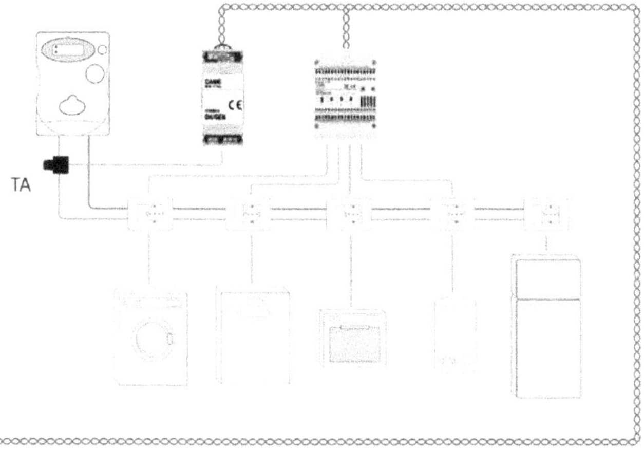

You can use more than one TA, for example by connecting one TA for each controlled outlet and thus have the power monitoring of each individual controlled appliance.

5.7.2 Lighting

Domotic also allows to manage lighting only when it is necessary, that is, depending on the presence of people and the amount of light in the room. A sensor can detect the movement of a person and check whether the amount of light, detected by a dedicated sensor, is insufficient (twilight function), therefore turn on the artificial lighting in the only period in which it is really needed.

5.7.3 Intrusion

By means of volumetric or magnetic sensors for detecting the opening of doors and windows, possible intrusions into the building can be checked. It is possible to protect every building and supervise it from the home system. In advanced systems, through appropriate communication interfaces, it is possible to monitor the map of the premises and, in case of alarm, to have the indication of the area from which the alarm originates.

5.7.4 Gas leak detectors

The gas leak detector is a sensor that detects the presence of gas in the environment, it must be connected to a dedicated module. When the gas concentration exceeds a specified threshold, the system shall signal the alarm in an optical and/or acoustic manner. It is possible to control the closure of an electro valve to block the gas release, or to interrupt the electricity to secure the environment.

The sensor shall be installed approximately 30 cm from the ceiling in the case of methane gas, or 30 cm from the floor in the case of LPG gas.

5.7.5   Water leak detectors

The water leakage detector is a sensor that allows, by means of a probe to be installed on the floor, to detect the beginning of a flooding resulting from a water leak. The sensor must be connected to a dedicated module.

When the water reaches the probe, the system signals the alarm optically and/or acoustically. A water supply closure solenoid may be operated.

5.7.6   Thermoregulation and air conditioning

Several operations can be carried out by means of a temperature sensor, connected to the domestic installation:

- Turn the heat on and off, divided by zones.
- Set the season (Summer/Winter).
- Set the desired ambient temperature.
- Set the maximum speed of any ventil convectors to improve the comfort of the living.
- Display the open/closed state of any window in the affected area.
- Lock the ambient thermostat keyboard to protect its setting.

6   CONNECTION TECHNIQUES

Now that we have some more ideas about what type of sensor, we can use for our purposes it is time to install them on our equipment, but above all we must electrically connect them in the right way!

6.1   Sensors with ON-OFF OUTPUT

Almost all the sensors come with the ability to be connected to an electronic board as well as to an industrial PLC, through a supplied cable, including the number of conductors needed.

The sensor can be an integral part of the sensor, then it is called sensor with cable output, or the sensor can have a connector to which connect the electrical cable, sold separately, which in turn has a connector. This second solution is slightly more costly but much more practical both during installation and in the case of subsequent maintenance, as to replace a possible broken sensor, simply remove the connector without having to disconnect the entire cable from the system.

Sensor connectors are usually circular with a screw-on lock ring. The most commonly used connectors are M8 or M12, with 2- to 4-conductor cables. Some special sensors may have cables with 8 conductors or more while maintaining the connection with the M12 connector.

The male connector is attached to the sensor, while the cable has the female connector, so it is not possible to accidentally touch the contacts of the connector, which has the cable connected to the electrical panel, causing dangerous short circuit. Sensor connector cables are equivalent between brands. Very small sensors may have a cable output with the M8 connector located along the cable.

Most commonly used sensors have 2- or 3-wire connections. For a 2-wire sensor these are usually BLUE and BROWN while for 3-wire sensors they are BLUE, BROWN and BLACK.

Be careful when connecting sensors to 2 and 3 wires. In a PNP system the 2-wire sensor shall be connected with the BROWN wire to the positive feed (24V) and the BLUE wire shall be its output (generally connected to the input of a PLC). For the 3-wire sensor, however, the BROWN wire is connected to the positive 24V, the BLUE wire is connected to the negative power supply, while the BLACK wire is the output (to the PLC).

For sensors with more wires, see manufacturer's specifications. Generally, a fourth wire can be used if the sensor has two outputs at the same time, one inverted to the other, or for learning functions with automatic sensor calibrations performed by the logic of the system without the intervention of a technician. Sensors may also exist with both ON-OFF and analog outputs that can be picked up via the same cable.

There are sensors that have the ability to work at network voltage (230 Vac). Be very careful when connecting these types of sensors.

6.1.1   Two-wire proximity sensors

There are two-wire sensors, in which a small current is circulating to keep the circuit running even when the sensor is not being read (OFF state).

Due to this phenomenon, it can happen that it remains on the load, if directly connected to the sensor output as shown in the figure, a small voltage that prevents the load itself from resetting. The problem may arise, for example, by using as load a small relay that can no longer be released once activated. Check the sensor and load technical sheets in these cases to prevent malfunctions.

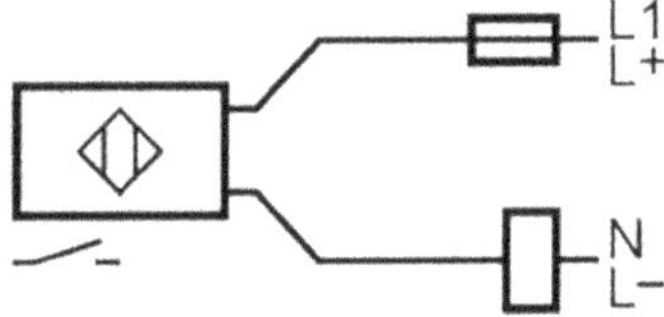

When the load current, at the active sensor, is less than a limit value, around 5 [mA], it is necessary to increase this current by placing a suitable value resistance in parallel to the load so as to provide the sensor with the minimum current for proper operation. Always consult the sensor documentation.

6.2   ANALOG OUTPUT transducers

A typical analog output transducer may present itself with a cable output or connector, as in the ON-OFF sensors.

Usually, the transducers of this type are connected individually to the control device, which can be an analog module of a PLC, a thermoregulator or a dedicated electronic board.

For transducers with a voltage output the most common type of connection is with a three-wire cable, power supply + e - and output signal, and it is advisable to use a shielded cable to avoid having electrical disturbances that affect the output signal.

For transducers with a current output, it is possible to have a cable with 2 or 4 conductors, 2 for the power supply and 2 for the current loop.

The more advanced components can also have a serial link (RS232 or RS485 standard) that allows direct connection to a PC or a PLC module.

6.2.1   Thermocouples and thermoresistance

These types **of temperature transducers** often have a connection with a two-wire cable. In almost all cases, a connector is not provided, and the electrical cable is an integral part of the device. PT100 thermoresistance can have 3 or 4 wires.

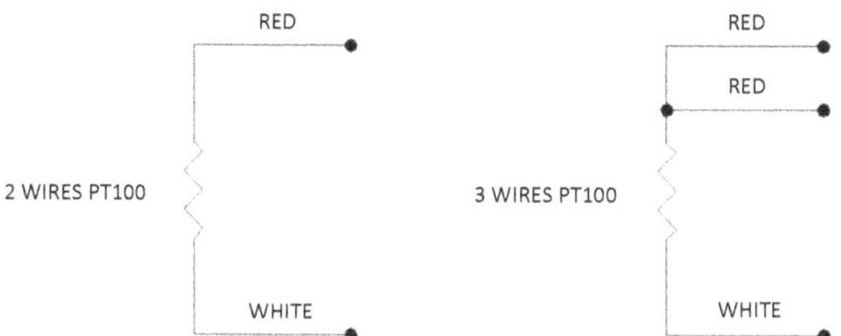

The **thermocouples** are polarized, so the polarity in the link must be respected. The colors of the wires may vary depending on the type of thermocouple, moreover the same type of sensor may have a different color depending on the nationality of the manufacturer.

The thermocouple wires are made of the same material as the joint. **Do not extend the wires if you don't need to**. If there is no alternative to procure the appropriate cable, consult the manufacturer.

The following table shows the coloring of the wires according to IEC 60584-3 international standard, for extension and compensation cables see also DIN EN 60584-1 standard. White wire is always negative.

The coloring of the wires used in the United States is completely different, in this case the orange color wire is always the negative pole.

	THERMOCOUPLES				
	J	**K**	**AND**	**T**	**R**
Materials	**Fe-CuNi**	**NiCr-NiAl**	**NiCr-CuNi**	**Cu-CuNi**	**Cu-CuNi**
Positive Thread	Black	Green	Purple	Brown	Orange
Negative wire	White	White	White	White	White

The most common industrial PT100 **thermoresistance** type are 2-wire and 3-wire thermoresistors, while the 3-wire thermoresistance allows a more accurate measurement because it compensates for the errors introduced by the resistance of the connecting wires themselves.

PT100s in a 4-wire configuration are often used in laboratory applications for greater precision. With four-wire connection, measurement is not affected by the resistance of the wires themselves or by the ambient temperature at which the wires are subjected.

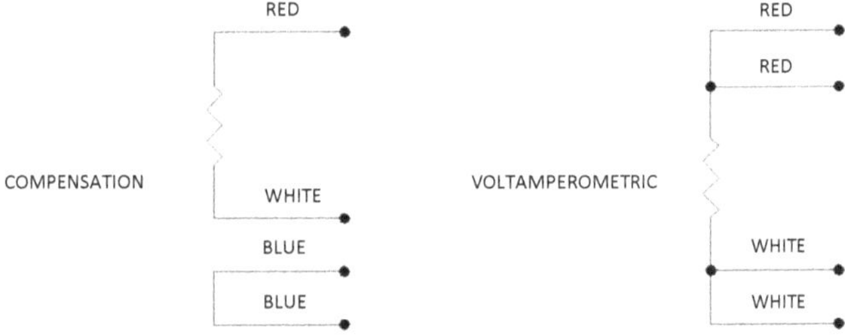

The compensation linking technique has now fallen out of use, classical voltamperometric is used.

In two-wire models, cable resistance can be compensated by external compensation resistance. Refer to manufacturer datasheets.

6.2.2   Load cells and bridge transducers

The following is a classic link of a Wheatstone bridge transducer. Two wires are used for the supply and the remaining two for the signal to be sent to the control electronics.

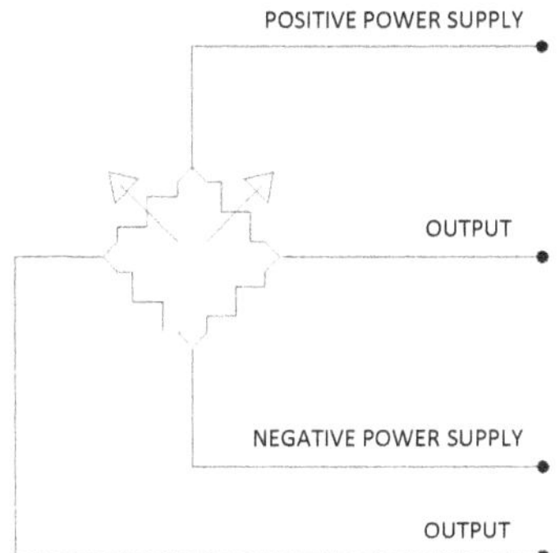

Often these types of transducers are sold with a packaging card supplied by the same manufacturer.

Wire colorings can change depending on the manufacturer and model, always check the catalog and technical sheet in the sensor before making the links.

6.3   ENCODER

Encoders, as seen in chapters 4.1.2 and 4.1.3 above, always require power supply for operation, so be very careful about the connection. Many encoders run at lower voltages than the standard 24Vdc (12V or even 5V), making sure the correct supply voltage is available. Also verify that the encoder output itself is compatible with the input circuit: motor board, dedicated electronics, PLC...

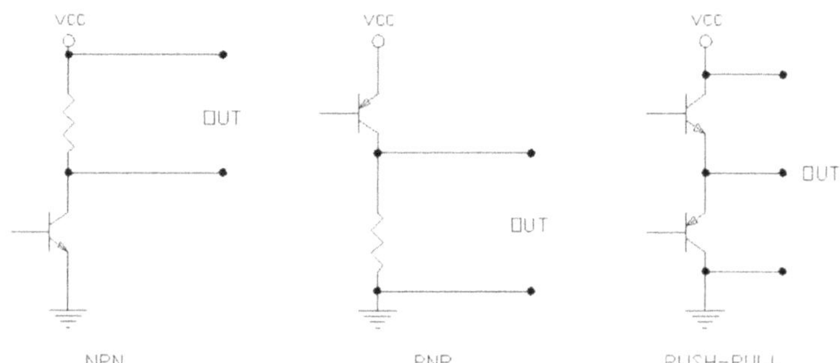

A typical encoder with channels A, B, and Z has a power cord with the wires Red (positive), Black (negative), and other colors, typically White, Yellow, and Green for the various signals.

Always check the encoder technical sheet and also the label on the encoder. Usually, the link wires with color and function are indicated on the label. The supply voltage value shall also be indicated on the label.

6.4   Point to point and field BUS

Once we choose the type of sensor and connecting cable, we finally have to connect it to our system. Rarely will we have a single sensor to connect, much more likely will have a number of different sensors to connect together with the central logic of the system.

Suppose we have a PLC in our electrical box that will have to manage the logic of the equipment we are setting up. We can connect each sensor to a different input of the PLC by arriving directly with the sensor cable to the PLC, passing, if necessary, into a terminal block if the sensor cable is too short.

This connection is called **point to point**, each sensor is individually connected to the logic of the installation. So, one of these is the cheapest connection in terms of cost, although impractical if we had a lot of sensors in the same plant.

In the case of many sensor systems, there may be hundreds in a large automatic machine, it is preferred a **field bus** connection.

A field bus is a means of connecting multiple sensors, not just sensors, to local terminals, with a traditional terminal or with more practical connectors. These terminals, usually 8- or 16-seat modules, are then connected to each other by a special cable that connects to the PLC, either directly or via an optional module. In practice the system is like a small network with serial connection, in which the PLC controls the status of the sensors by dialoguing with the modules on the plant, near the sensors themselves. This drastically reduces the number of wires from the system to the switchboard, as **only the BUS cable** connecting the PLC to all modules is sufficient.

There are several different field bus types, but the connection is similar for all. The industry standards are different and not compatible with each other, the modules to which the sensors connect have an electronic board specifically designed for only one type of field bus, so you have to make sure that all the modules you intend to connect, and the PLC are expressly built for the bus type you want to use. The most common standards include ProfiBus, DeviceNet and CANBus.

Field bus connection is much more **expensive** than point-to-point, we can consider broadly a cost 5-10 times higher, but the initial investment pays off as the connection of the various sensors is much faster and many fewer wires are used. In addition to this, one has the advantage of having a more compact PLC as it takes only **one** communication **module** with the field bus to check the status of **all the sensors**, thus being able to realize an electrical picture of smaller size.

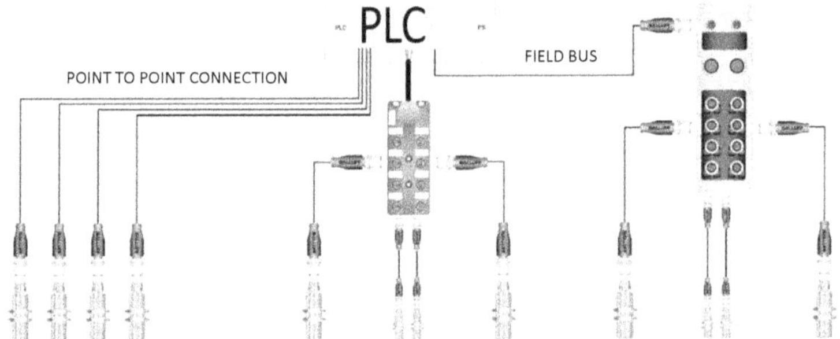

See the example of a traditional link for the four sensors on the left, all connected via connectors. Each sensor reaches the PLC inlet with its own cable.

The other links are made by field bus and modules to connect up to eight sensors each. The module in the center connects to the PLC via a single cable, while the module in the right has connectors for cascading other similar modules via the bus. In any case, the PLC comes with **a single cable** with considerable savings in wiring time and space used.

6.5   Parallel series of ON-OFF sensors

Many manufacturers publish in their catalogs and sensor data sheets connection examples of two or more sensors in series or in parallel.

In the series connection, the first active sensor feeds the second, the output of which goes to the control circuit.

In the parallel connection, the output of multiple sensors goes to the same input as the control circuit.

Some examples of links are given below. At the top is a series link for PNP and NPN sensors, at the bottom are the links to be made for a parallel configuration.

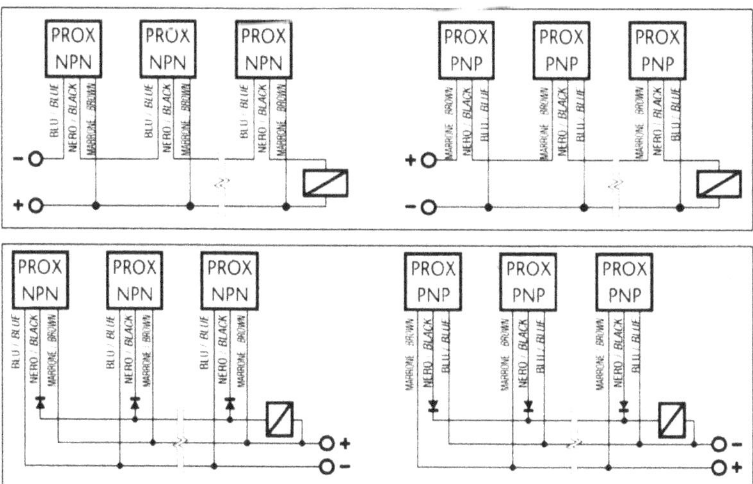

Note in the image the diodes used in parallel connections on the sensor output wire. Use common silicon diodes type 1N4148. For convenience, the load is referred to as a relay coil but could be an input to a PLC or other device.

By connecting sensors in **series**, the logical **AND** function is realized, that is, if all sensors are read, I have an output signal. By connecting sensors in **parallel**, the **OR** logic function is realized, a single read sensor is sufficient to have the signal output.

If it is absolutely necessary to make a SERIES or PARALLEL connection, follow carefully the manufacturer's instructions of the sensors used.

Problems can arise in the search for faults, in the correct positioning and wiring of sensors in series or PARALLEL, this practice is not recommended if not necessary. Connect each sensor to its own PLC or control board input.

6.6 Cables and connectors

Sensors and transducers designed to be connected via a connector cable usually have an H1 type connector with M8 thread, a H type connector with M12 thread, or a K type central attachment screw connector.

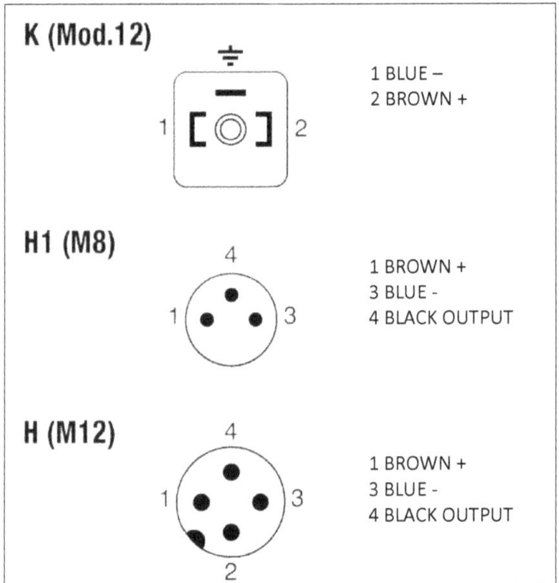

The connectors with the pinout and cable colorings that are usually used are shown in the figure. The view is of the male connector, which is usually incorporated in the sensor.

In the figure below a typical multi-pole cable with H-type connector is shown, in this case a five-pole M12 connector is visible, often the central pole is not used.

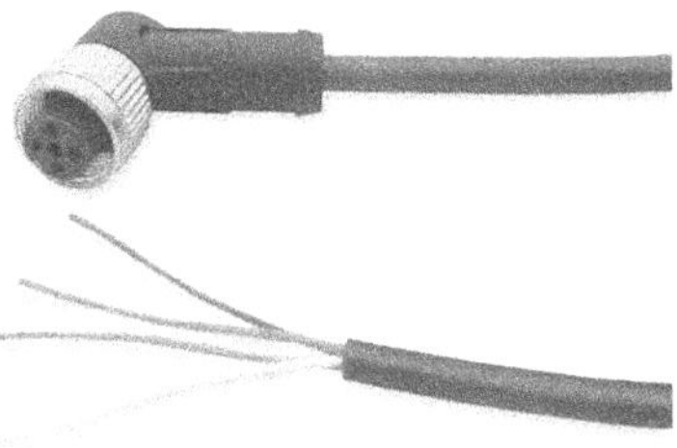

Refer to the documentation of the sensor used to make the correct connection

We have seen that there are different types of sensors that can be classified according to the size to be measured, the principle of operation and also according to the type of electrical output implemented.

A sensor can be found for virtually any purpose, for detecting objects of all types and materials, including transparent materials, liquids or dust; for each type of measurable size such as pressure, temperature, force, speed or acceleration, and also for special uses such as color detection, barcode readings, and control of object shape and size.

This chapter summarizes criteria and suggestions for choosing the appropriate sensor depending on the intended use.

In order to speed up the choice of a sensor, given the large number of models and technologies used, it is appropriate to ask a few questions before starting the search for the component that best satisfies our application. The following tables may be used as guides.

7.1.1   Preliminary Sensor Selection Assessments

Analysis and considerations	Notes	Usable Sensor
– Is this a new application? – Is a sensor already being used for the same purpose? – Is the current operation unsatisfactory?	With a sensor of the same type, even if of another brand, the same problem could arise	-
– How is the object to be detected?	Evaluate COLOR, SHAPE, SIZE, MATERIAL	-
– What distance is the object to be detected?	-	For short distances, some millimeters, orient towards inductive or capacitive sensors, otherwise it is better to adopt optical or ultrasonic sensors
– Supply voltage	Check the voltage available in the electrical system: if DC or AC and its value	-

Analysis and considerations	Notes	Usable Sensor
– Sensor load – Sensor output connection	Assess the load to be flown by the sensor. Will the sensor be connected to a PLC or card, or should it drive a relay, contactor or other loads? Check if the sensor has to be PNP or NPN output	-
– Mounting	Evaluate, in particular in the selection of inductive or capacitive sensors, the possible proximity of objects that could disturb the sensor, including the sensor housing	-
– Without the piece to detect what is in front of the sensor? – If you don't have the piece you want to detect, is there anything that can get into the field of reading?	Evaluate background distance, color, shape, and material	For problems with surfaces behind the piece to be detected, use a background suppression or refracting sensor with refraction
– Is the object being detected moving?	Calculate at what speed the object passes in front of the sensor and how often	A very fast object can be detected, but it may be necessary to adopt a delayed-turn sensor to make sure that the sensor output signal is recognized without uncertainties by the control system (PLC, industrial PC, dedicated instrument, etc.).
– Should the output signal of the sensor be on with or without the part to be detected?	-	Choice of normally open or closed sensors (active without the target), or light-on or dark-on optical sensors (active with receiving radius or interrupted)
– Environment	Evaluate the environment in which the sensor will have to work: TEMPERATURE, DIRT, VIBRATIONS, HUMIDITY	If necessary, refer to "IP Security Rates" in the APPENDICES

7.1.2 Assessments of object size to be detected

Size	Detection only (on-off)	Control (analog output)
DISTANCE POSITION	– Optical Sensor – Ultrasonic Sensor	– Potentiometer – LVDT transducer – Encoder – Hall effect sensor – Optical transducer

SPEED	-	– Ultrasonic transducer – Magneto strictive transducer – Encoder – Resolver – Tachymetric dynamo
PROXIMITY	– Induction or capacitive sensor (few mm) – Microswitch (contact) – Optical Sensor – Magnetic Sensor	-
TEMPERATURE	– Thermostat	– Thermocouple – Thermoresistance
PRESSURE	– Pressure switch	– Pressure transducer
ACCELERATION	-	– LVDT – Extensimeter – Accelerometer
FORCE MASS	-	– Extensometer (load cell)

7.1.3 Evaluations by application type

Application	Object or size to be detected	Usable sensors
DETECTION	TRANSPARENT OBJECTS – plastic packaging – glass or PET containers	– Capacitive – Optical fork or measuring – Ultrasonic Sensor
DETECTION	HOLED OBJECTS – wire mesh – piece of machinery – printed circuit boards	– Viewing System – Ultrasonic Sensor – LASER – Optical fiber sensor
DETECTION	CONTRAST COLOR LUMINESCENCE SHINE	– Color Optical Sensor – Luminescence Sensor – Contrast or shine sensor – Viewing System
DETECTION	DIMENSIONS PROFILES	– Ultrasonic Sensor – Optical for Distance Detection – Measuring optics
DETECTION	METALLIC MATERIALS	– Induction Sensor
DETECTION	MAGNETIC MATERIALS – permanent magnets – stems of pneumatic actuators	– Magnetic Sensor – Magnetic sensor or reed shaped for pneumatic actuators
DETECTION	MARKINGS LABELS	– Color Sensor – Contrast Sensor – Viewing System – Barcode reader or arrays – Reflection optical sensor
DETECTION	TRANSIT BY TRANSPORT	– Rear-Reflection Sensor – Barrage Sensor – Optical fiber sensor – Optical barrier – Optical fork sensor – Optical Sensor with measuring function
DETECTION	DOUBLE LAYER MATERIALS – paper or metal stacked sheets – transparent film	– Fork or optical fiber sensor with sensitivity or self-tuning adjustment – Ultrasonic Sensor
DETECTION	EDGES – moving object edge	– Optical Sensor – Optical barrier – Measuring LASER Sensor – RADAR sensor (for outdoor applications)
DETECTION	CONTAINERS – full/empty detection – detecting parts in a container	– Ultrasonic Sensor – Optical distance sensor – Viewing System

Application	Object or size to be detected	Usable sensors
MEASUREMENT OR CONTROL	DIMENSIONS PROFILES VOLUMES	– Viewing System – Optical Sensor with measuring functions – Encoders with special applications of measuring wheels – Optical barrier
MEASUREMENT OR CONTROL	TEMPERATURE	– Temperature transducer – Thermostat
MEASUREMENT OR CONTROL	PRESSURE	– Pressure transducer – Pressure switch
MEASUREMENT OR CONTROL	LEVEL – level of liquids, fluids, dust	– Ultrasonic Sensor – Capacitive Sensor – Optical fiber sensor – Level sensor (Chapter 3.7) – Encoders with appropriate mechanical accessories
MEASUREMENT OR CONTROL	DISTANCE	– Magnetic Position Transducer – Optical sensor for measurement or distance – Ultrasonic Sensor
MEASUREMENT OR CONTROL	SPEED	– Encoder – Resolver – Tachymetric dynamo
MEASUREMENT OR CONTROL	DUST OR GAS CONCENTRATION	– Gas Sensor
MEASUREMENT OR CONTROL	FLOW SCOPE	– Flow or flow transducer
MONITORING	EMISSIONS	– Gas Sensor
MONITORING	QUALITY	– Viewing System – Color, luminescence or contrast optical sensor – Barrage Sensor with measuring functions
MONITORING	PROCESSES	– Flow Transducer – Pressure transducer – Level Sensor – Temperature transducer – Gas Sensor
SECURITY	HAZARDOUS POINTS points inside machines	– Optical Security Barrier – LASER Scanner – Safety RADAR Sensor – Sensitive safety carpet

Application	Object or size to be detected	Usable sensors
SECURITY	ACCESSES access to areas with hazards or within installations	– Optical Security Barrier – LASER Scanner – Safety RADAR Sensor – Safety Microswitch for Rear Aperture Control – Sensitive safety carpet
SECURITY	DANGEROUS AREAS unfenced areas	– Optical Barrier or Security Camera – LASER Scanner – Safety RADAR Sensor
SECURITY	DOORS AND SHUTTERS verifying the position of mechanical access control elements	– Rotary or linear security encoder – Security Microswitch with or without contact
SECURITY	MOVEMENT Hazardous Parts Rotary Motion Detection	– Rotary or linear encoder – Tilt Sensor, Tiltmeter
POSITIONING	POSITION RECOGNITION	– Encoder – Viewing System – Optical Sensor for Measurement – Distance Sensor – Ultrasonic Sensor – Inclinometer
POSITIONING	EDGE GUIDE edge detection or contrast changes	– Viewing System – Contrast Sensor

7.1.4 Practical examples of applications

Plastic bottle filling control with a capacitive sensor:

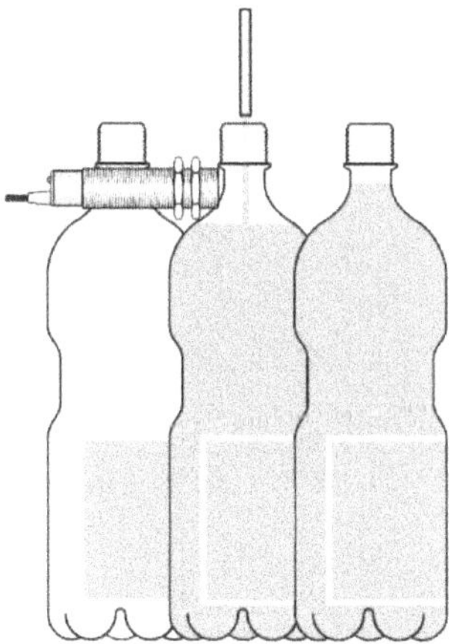

Level control of a liquid or powder tank with capacitive operation probes:

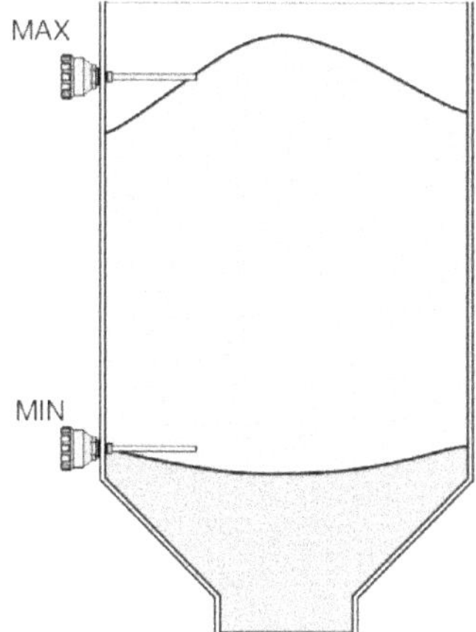

Formations of probe deposits can distort or prevent measurement, the use of probes covered with non-stick materials is recommended. Materials with very small dielectric constants or near 1 may be difficult to detect

Control of the presence of the cap on bottles with an optical LASER sensor with refraction:

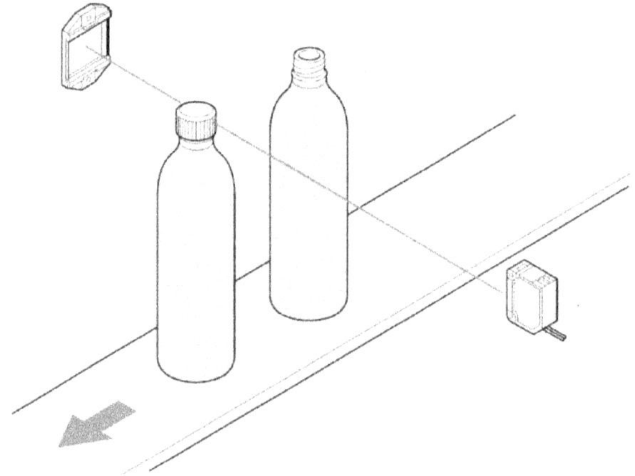

Triggers for the vision system when passing parts on tape made with an optical LASER sensor:

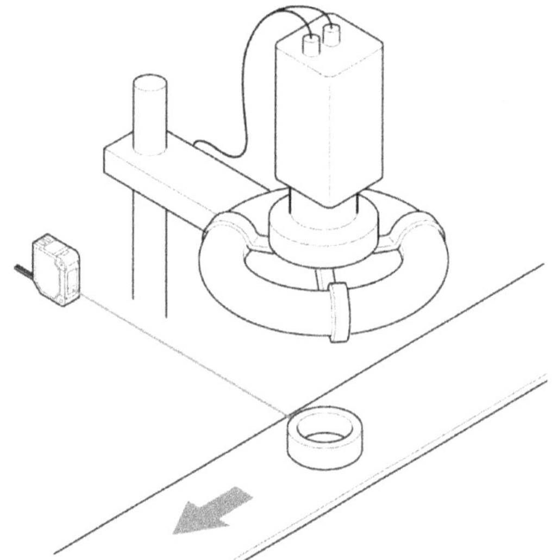

Verification of the length, when passing, of hexagonal head screws by means of a visible light optical sensor and fiber optic barrage. Only the emitter is visible in the figure:

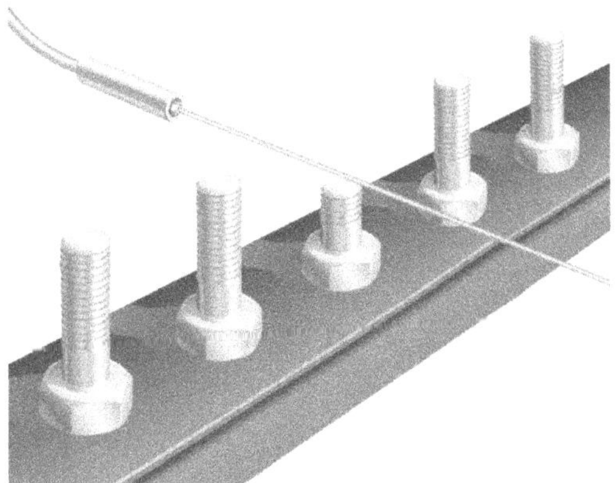

Control of printed circuits manufacturing, check the solder layer, using an optical sensor with precision measurement functions:

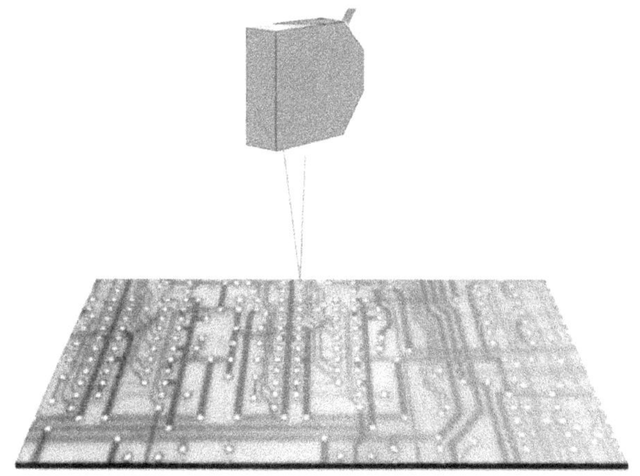

The following tables indicate the common acronym and graphic sign used to indicate the different sensors in an electrical scheme. Given the wide variety of sensors and transducers, it is not always easy to identify the correct symbol and its acronym.

For graphic signs, it may be useful to consult EN 60617 in case of doubt. Many CAD programs for electrical plant design have incorporated very comprehensive symbol libraries that comply with current regulations.

The catalogs of the various manufacturers may indicate new or different symbols than those shown without this being a problem, the important thing is to identify the component drawn unequivocally.

The following is a table for help in choosing the correct symbol to associate with the graphic symbol depending on the type of sensor. In case of doubt, only the first letter (B, S) can be used. In extreme cases, the letter E can be used to represent any type of component that cannot be placed in a well-identified category.

Acronyms that can be used in an electrical scheme			
First letter	*Identification*	*Sensor type / Detected Size*	*Complete code*
B	Transducers of a non-electric magnitude in an electric unit and vice versa Analog output	Pressure transducer	BP
		Position transducer	BQ
		Tachymetric dynamo	BR
		Temperature transducer	BT
		Speed transducer	BV
S	Control and control equipment Sensors with ON-OFF output	Liquid level sensor	SL
		Pressure switch	SP
		Proximity Sensor	SQ
		Rotation Sensor	SR
		Thermostat	ST

In the following table the symbols usually used for some sensors.

Symbols usable in an electrical scheme		
Type	**Symbol**	**Notes**
Inductive Sensor	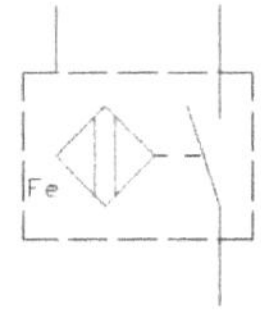	The wire attached to the dashed container, for PNP sensors is the negative, 0 V
Capacitive Sensor	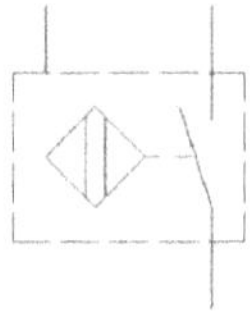	The wire attached to the dashed container, for PNP sensors is the negative, 0 V There are sensors with amplifier separate from the read head, the symbol remains the same
Barrier optical sensor	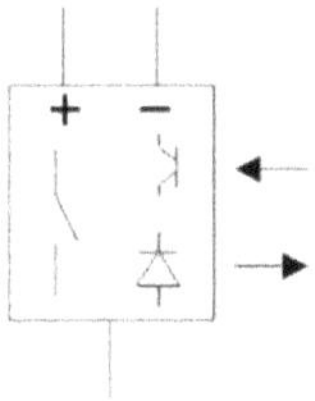	This type of sensor consists of two distinct elements: projector and receiver. A unique symbol is used in the drawing
Optical retroreflection sensor	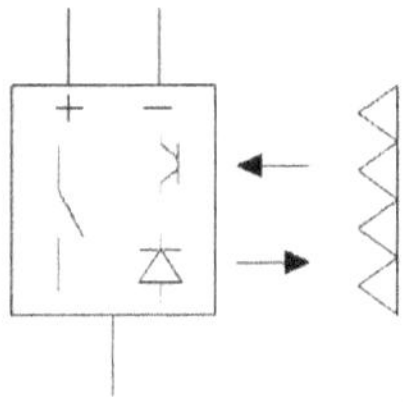	The graphic symbol shown here is composed of the symbol of a barrage sensor plus the right-hand part representing the refracting. It is generally accepted in this form
Direct touch optical sensor	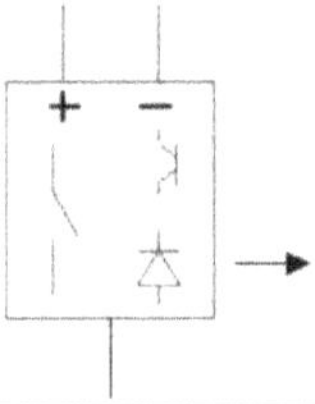	-

Symbols usable in an electrical scheme		
Type	**Symbol**	**Notes**
Optical Contrast Sensor	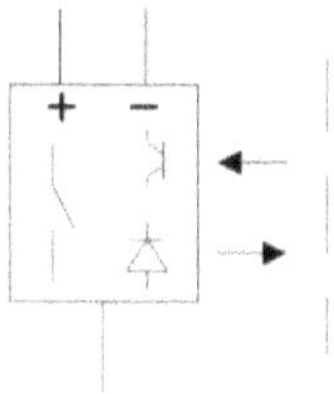	-
Optical Luminescence Sensor	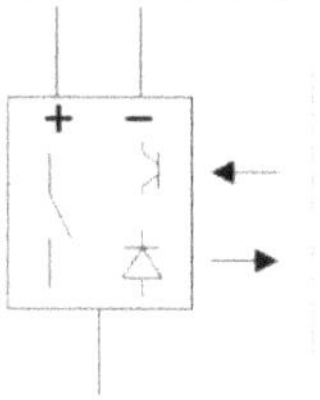	-
Optical RGB Color Sensor	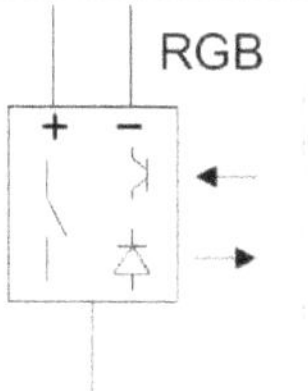	-
Optical distance sensor	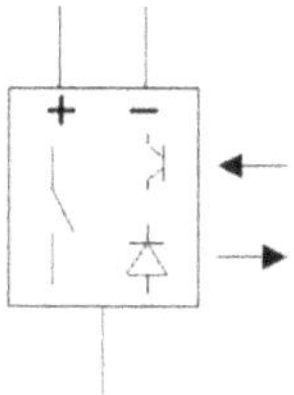	The symbol is identical to that used for the barrage sensor but in this case the sensor consists of only one element
Mechanical Microswitch	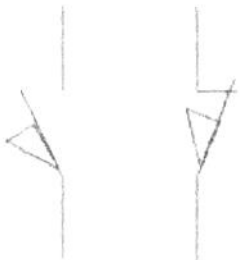	Here is a normal open contact on the left, the normally closed contact on the right.
Magnetic Sensor	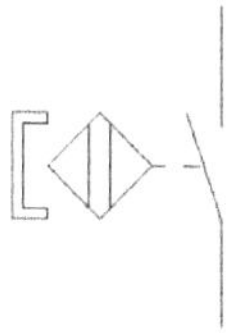	-

A mechanical switch may also have multiple contacts operating at the same time. Various combinations may be present, including mixed NA and NC contacts. In the case of multiple contacts, it is good to combine with a hatch, in the diagram, the symbols of the various contacts that belong to the same microswitch

9   APPENDICES

9.1   Explosive Atmospheres - ATEX

Potentially explosive atmospheres are all those places, identified and classified, where the danger of an explosion may exist. The classification of potentially explosive zones is a preliminary assessment that allows the risk assessment and the selection of components appropriate for use in the various zones.

The zones are classified according to the type of explosive atmosphere present, gases or dust, and the frequency and duration of the atmosphere.

Many seemingly harmless materials can become potentially explosive, when ground or powdered and when mixed in the right proportions to air. The cause of the explosion is called the ignition source, it can be a spark due to the closure of an electrical contact, a component with a sufficient surface temperature, mechanical friction, etc...

To limit the risk of explosion, in areas classified as explosive, compliant components, and thus sensors, must be used. The directive governing components is Directive 2014/34/EU. The sensors which may be used shall bear the symbol:

In addition to marking, the documentation of the sensor shall include a number of indications identifying in which atmosphere and situations it can be used.

9.1.1 Classification of zones

ZONE	MATERIAL	EXPLANATION	USABLE SENSOR
0 20	GAS DUST	place where an explosive atmosphere consisting of a mixture of air and flammable substances in the form of gas, steam or mist is present continuously, or for long periods, or frequently	Category 1
3 21	GAS DUST	Place where an explosive atmosphere, consisting of a mixture of air and flammable substances in the form of gas, steam or fog, is likely to occur occasionally during normal operation	Category 2
2 22	GAS DUST	Place where an explosive atmosphere consisting of a mixture of air and flammable substances in the form of gas, steam or fog is unlikely to occur during normal operation, but which, if present, persists only for a short period	Category 3

9.1.2   Sensor marking

The sensor shall be marked according to the provisions of ATEX Directive 2014/34/EU, see an example of a possible marking with mandatory supplementary indications:

$$\text{C}\,\text{E}\ \langle\text{Ex}\rangle\ \text{II 2G IIB T6}$$

The meaning of the marking may be summarized by considering the individual indications:

- CE marking $\text{C}\,\text{E}$ and ATEX marking $\langle\text{Ex}\rangle$.
- **II** is the set of devices. Group I is for use in mines, Group II is for use outside mines, i.e., at the surface for all environments.
- **2** indicates the category of membership: category 1, 2 or 3 depending on the increasing risk.
- **G** or **D** indicates whether the sensor is suitable for use with gas (G) or dust (D).
- **IIB** denotes the group of gases for which the sensor can be used, in this case IIB denotes ethylene-type gas compatibility. The gases are divided into four groups (I, IIA, IIB, IIC). In the first group we find methane, in the last we find hydrogen.
- **T6** indicates the temperature class, from T1 corresponding to use up to 450°C, to T6 for use at 85°C maximum. Indicates the maximum surface temperature that must not be exceeded, otherwise triggers may occur.

There may be additional indications, such as the protection factor or protection modes. Always consult all documentation provided by the sensor manufacturer. Always seek expert advice before using equipment in potentially explosive environments.

9.2 Safety applications

The safety of machinery is often dependent on the design of measures to prevent or control access to or the presence of persons in dangerous areas. The hazards can be of various kinds: moving mechanical parts, high temperature surfaces, radiation emitting zones of various types.

It is important, therefore, to protect people who have to work in contact with machines. To this end, there are specific rules describing how protections (both openable and stationary), and sensors to be used for the safety of persons, must be made. Security sensors can be of various types, from classical mechanical micro-switch to modern RFID sensors to volumetric LASER scanners.

9.2.1 Classification of safety sensors

ISO 14119 standard deals with the requirements criteria for choosing and applying safety sensors, and also classifies safety sensors according to the type and level of encoding implemented.

The encoding of a safety sensor is to prevent access to the machine under abnormal conditions, such as when using a "universal" actuator key to be inserted into the mechanical switch to work with an open protection door.

Type	Pickup device	Detection Principle
Type 1	Mechanical uncoded	Contact, force. Microswitch lever or hinge
Type 2	Mechanical coded	Contact, come on. Keyboard microswitches (with shaped actuator), trapped key
Type 3	Contactless uncoded	Induction, capacitive, magnetic, ultrasonic, optical
Type 4	Contactless coded	Magnetic sensors, RFID, optics. With encoded magnet, RFID encoded or optical encoding

The choice of sensor type to be used, the encoding level and any measures to be taken against the tampering of the sensor shall be made after a specific risk analysis for the machine being treated. There is no sensor suitable for all uses, the choice has to be assessed case by case.

For the verification of the safety features implemented, such as the shutdown of an engine when a door is opened, a specific standard such as ISO 13849-1 is required.

9.2.2　Microswitches

Among the sensors still most used today we have mechanical micro-switches, often used to control the opening of movable shelters, doors and protective casings.

A particular and widely used version of these microswitches has a separate actuator, commonly called a key, which enters the microswitch body to drive the contact. In the figure is shown a microswitch with a separate key actuator, in this case the key is encoded, as seen by the shaping of the end of the actuator, which reminds the profile of a common household key for the opening of a door, each microswitch must be actuated from its own key making it unlikely to use a universal key to actuate the machine with open repair.

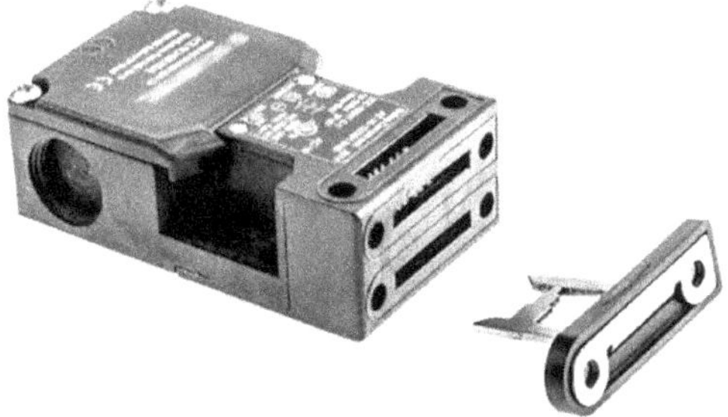

These special microswitches consist of a body shaped like a parallelepiped that houses the electrical contacts and a fork shaped key that inserts inside the body under normal working conditions. The body may be attached to the frame of the machine and the key to a protective door in such a way that the opening of the door itself causes the key to be released from the microswitch and the activation of the safety function resulting in the stopping of dangerous movements.

Use only certificate sensors for safety applications, combined with certificate control modules. Ensure that all components and their connection comply with the manufacturers' requirements. A sensor for general use should not be used for safety applications.

The safety toggles may also have the mechanical lock restraint function of the key, to prevent the opening of the protective cover. This function is implemented with a solenoid that activates a mechanical latch, the solenoid needs to be fed separately, and can be controlled by special safety modules. Different vendors offer a variety of solutions for a variety of applications.

The peak current at solenoid activation is very high. Check the power supply capacity and consult the microswitch manufacturer catalogs that you intend to use. Some microswitch models are available with delayed or staggered solenoid insertion, when multiple microswitches are used at once.

9.2.3 Magnetic sensors

Specific magnetic sensors exist for safety applications. These sensors must be combined exclusively with their magnets, which have magnetic field coding and alternations. This means that a normal magnet cannot be used to activate the sensor and thereby evade the safety function. The following are examples of magnetic sensors with paired magnet. The sensor at the top has a connector for electrical connection, the one at the bottom has a built-in cable.

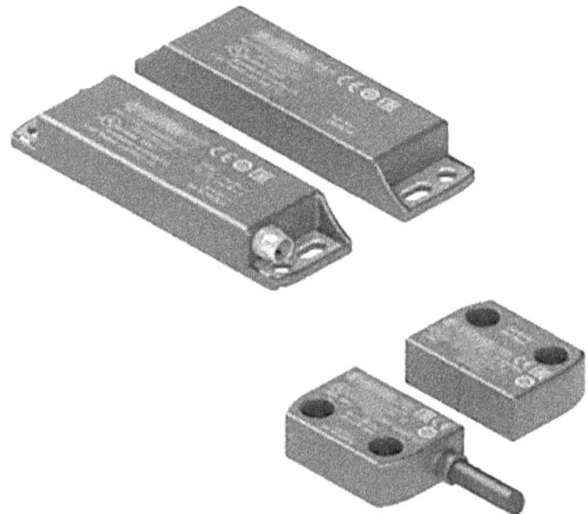

Modern magnetic sensors can be connected in series with each other, maintaining the category and reliability stated by the manufacturer, thanks to internal diagnostic and control functions, thus exceeding the great limit of mechanical microswitches.

9.2.4 RFID Sensors

Similar in appearance to magnetic microswitches, they are based on Radio-Frequency IDentification (RFID) technology.

RFID technology allows data to be stored and read in devices called tags or transponders, which can respond to remote interrogation by dedicated fixed or portable devices called readers, in this case our safety sensor.

Identification of the tag is by radio frequency, which enables the sensor to read the information contained in the tag it is interrogating.

The tag is powered, at the time of reading, by the energy emitted by the radio frequency sensor. This is the same technology you see in store anti-dumping systems. You will have happened to buy an object that has a thicker label than usual with a spiral or a similar shape at its core. The spiral is the antenna that activates the circuit when passing through the outgoing barrier of the store. When you pay for the item, the cashier is written "paid" within the tag, using the radio transmission, so that you do not sound the alarm when you leave.

Sensor-tag pairs can be made with encoding, so that each sensor only functions at its own tag. This type of sensor can be equipped with an output dedicated to diagnostic functions, for reporting any problems or failures in the electrical connection or in reading the tag, for example due to an excessive distance between the sensor and the tag.

9.2.5   Optical barriers

An optical barrier is essentially a multi-beam optical sensor consisting of an emitter and receiver, usually housed in separate parallelepiped containers. The components of the barrier attach to the sides of the danger zone, passing an object interrupt one or more of the rays, triggering the safety function.

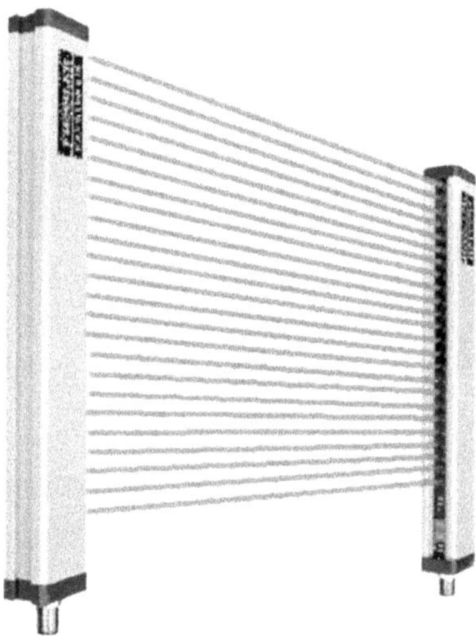

In practice, it is like having many photocells aligned to control access to a dangerous area: presses, moving organs, areas at risk of cutting or crushing.

A key feature of barriers is their resolution, in effect the space between rays, in which a finger, hand or arm could penetrate without activating the safety function. Barriers may have varying levels of protection, depending on their resolution and hence the size of the object causing the alarm to be triggered, and are usually referred to as finger, hand, arm or person (body) protection barriers.

Barriers can have several options such as suppression of certain rays or MUTING, the temporary deactivation of the barrier itself for the programmed passage of an object, such as a carton on a conveyor belt. The muting is triggered by external photocells that, by reading the object in the right time sequence, activate the function.

There are cascading barriers to protect areas with irregular profiles. For protection of perimeters, you can use mirrors to reflect the rays from the projector to the receiver.

9.2.6 Sensitive borders and carpets

Sensitive carpets and borders are special safety sensors. They are made of elastic material, plastic or rubber and are pressure sensitive.

The carpet or border is a passive entity and therefore requires a control module to perform the safety function for which it is responsible. Stepping on a carpet or pushing on a sensitive border is equivalent to opening a contact and thereby activating the expected safety feature, usually used to stop dangerous movements.

In the figure is represented a sensitive edge consisting of a metal guide on which is fixed the sensitive rubber border. Below is visible the power cable that can be used to connect to the safety module that controls the activation of the edge:

Rugs can have special features such as anti-slip surfaces or resistance to corrosive substances.

The following figure shows a typical sensitive rubber carpet with embossed workmanship. The electrical cable for the signal output is visible in the upper right corner.

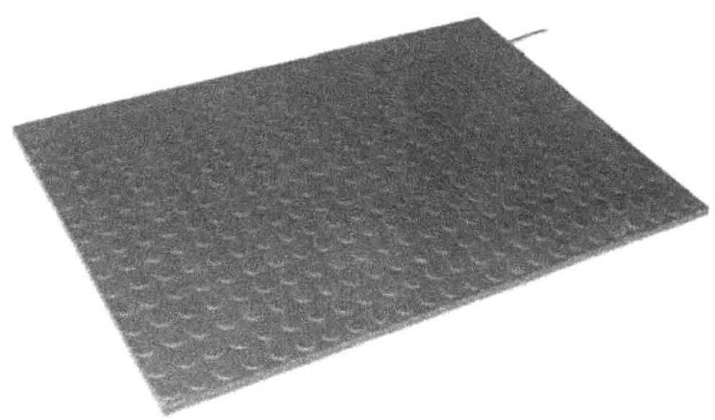

9.2.7 Volumetric sensors

The safety volumetric sensors commonly used are LASER scanners. They are thus optical sensors, intended for the surveillance of areas normally devoid of mechanical protection or other barriers.

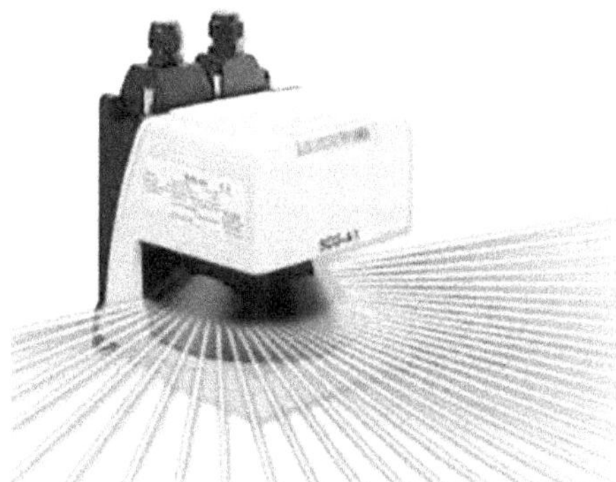

These scanners are programmable, to define intervention distances. The most advanced models can monitor an area with an irregular perimeter and have early warning outputs, in case of intrusion in the outermost supervised zone, and alarm outputs in case of intrusion in the inner zone. Programming of monitored areas, activated alarms and any other manufacturer's specified features is done using dedicated software.

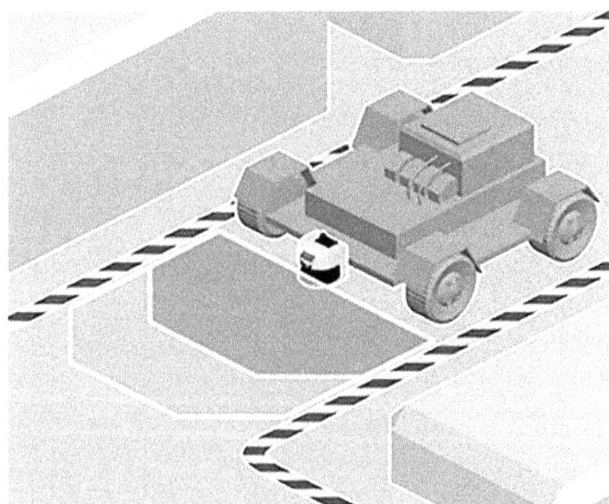

In addition to protecting people in areas without mechanical protection, scanners are often used on automated vehicles so that they can "see" an obstacle or person and brake, slow down or stop in time.

9.2.8 Radar safety sensors

They have recently appeared on the market for innovative sensors that use RADAR technology to detect access or presence of people.

RADAR technology is based on motion detection, so all objects that remain motionless are not seen. These sensors, therefore, unlike LASER scanners, do not need to be programmed to precisely define the perimeter of the area to be detected as much as an object, even if it is moved from its initial position, is no longer detected if it remains motionless.

These sensors are very useful for controlling the access of a person or object within the area being monitored.

They are also useful in giving consent to the restart of a car as they are able to check very precisely whether a person has remained within the perimeter of the machine. They succeed in this task thanks to their very high sensitivity, which is able to detect the simple breath of a person, even if the person tries to remain motionless.

This type of sensor also is not affected by variations in ambient light, dust, fumes and also resists in wet environments or with water splashes.

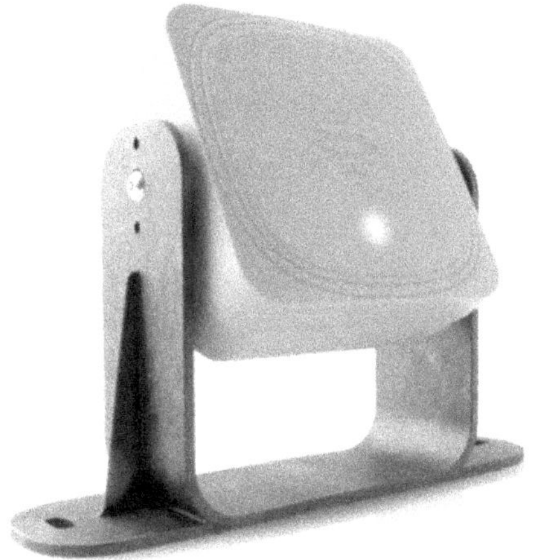

9.3   Degrees of equipment protection IP

Knowledge of the degree of protection, i.e., the ability of containers and devices to present a certain degree of resistance against the penetration of solid bodies, liquids or vapors, can help in choosing the correct sensor to use in different environments. Some installations are located outside, in explosive environments, in places with possible condensation or dust formation. You must choose a sensor that fits your location.

The IP designation indicates the degree of equipment protection required by IEC. The first digit indicates the degree of protection against access to foreign objects, the second digit indicates the degree of protection against water. The initials are IP-based and the digits, for example IP-54 can be used to identify a container that is protected against dust and splash.

Specific rules determine how to classify the degrees of protection of electric casings. CEI EN 60529 'Degrees of protection of casings' provides for a system of classification of the degrees of protection of casings for electrical equipment, to indicate the level of protection of casings against access to dangerous parts inside the casing and against penetration of foreign solid bodies or water. The standard does not consider protection against explosion risks or environmental situations such as humidity, corrosive vapors, molds or insects.

The declared IP degree of protection must be ensured in the 'ordinary service condition of the equipment'.

IP Code Structure:

1ᵃ digit	2ᵃ digit	3ᵃ digit	Additional letter	Additional letter
0...6	0...8	Ik00...IK10	A-D	H-W

The IP grade may only be indicated with the two characteristic digits, plus any additional letter, to indicate the degree of protection for persons against access to live parts, and an additional letter, to provide further specific product clarifications.

The degree of IP protection should always be read digit by digit and not globally. A designated envelope with a certain degree of protection shall also conform to the lowest degrees of protection, with the exception of the values 7 and 8 used for the second characteristic digit, which does not automatically lead to the fulfillment of the requirements for digits 5 or 6, unless it bears the double marking, for example IPX6/IPX7. A casing with degree of IP31 protection is suitable for an environment requiring a minimum degree of protection IP21. An enclosure with an IP30 rating may not be used in the same environment.

The first and second characteristic digits are mandatory. If the level of protection corresponding to one of the digits is not specified, because it is not necessary or because it is not known, it is replaced with an X. The additional letter and the additional letter are optional and can therefore be omitted without being replaced.

Only the installation, installation and maintenance carried out in accordance with the rules of the art shall ensure that the original degree of protection of a casing is maintained. In view of the fact that the presence of water on the equipment and pipes is in any case adversely affecting penetration, corrosive effects and so on, it is appropriate that the equipment installed on the outside be accompanied by a protective roof which may be complemented by side screens.

First characteristic digit

The first digit simultaneously indicates the protection of materials against the penetration of foreign solid bodies including dust, and the protection of persons against contact with dangerous parts, see also "additional letter".

1ª digit	Protected case against:
0	Unprotected
3	solid bodies larger than 50 [mm] and back-to-hand access
2	solid bodies larger than 12 [mm] and finger access
3	solid bodies larger than 2.5 [mm] and tool access
4	solid bodies greater than 1 [mm] and wire access
5	Powder and wire access
6	Totally protected against dust and wire access

IP1X protection is only permissible for equipment intended to be protected by a housing or installed in enclosed locations and accessible only to trained persons. IP2X and IP3X protection is provided for components installed in locations that are accessible to untrained users in day-to-day environments with small objects.

IP4X protection, which represents the highest degree of protection against the entry of solid bodies, is used when the presence of wires, chips, filings or other is expected.

The IP5X protection is suitable for occasionally dusty environments such as paved roads, steel plants, etc. The IP6X protection is suitable for permanently dusty environments such as cement or powder storage.

Second characteristic digit

The second number indicates the protection of materials against harmful penetration of water. The tests shall be carried out with fresh water without surfactants.

The figure IPX8 shall be supplemented with the indication of the maximum immersion depth which can be withheld from the container, as declared by the manufacturer.

2ª digit	Protected case against:
0	Unprotected
3	vertical drop of water drops
2	drop of water droplets with a maximum inclination of 15°
3	rain
4	splash of water
5	water jets
6	waves
7	effects of immersion
8	effects of submersion

Third characteristic digit.

The third digit indicates protection against mechanical impacts. It is usually omitted.

Digit	Protected container against:
IK00	unprotected
IK01	impact energy of 0,15 [J]
IK02	impact energy of 0,2 [J]
IK03	impact energy of 0,35 [J]
IK04	impact energy of 0,5 [J]
IK05	impact energy of 0,7 [J]
IK06	shock energy of 1 [J]
IK07	Impact energy of 2 [J]
IK08	Impact energy of 5 [J]
IK09	impact energy of 10 [J]
IK10	Impact energy of 20 [J]

Additional letter: protection of persons.

Additional letters shall be used where the protection of persons against contact with dangerous parts is greater than that of the entry of solid bodies expressed by the first characteristic digit. In other words, the letter, if any, indicates that the protection provided by a wrapper against access to dangerous parts is better than that indicated in the first digit.

This higher protection may be provided, for example, by barriers, openings of an appropriate shape or by internal distances to the casing.

The additional letter shall only be used if:
- effective protection against access to dangerous parts is higher than that indicated by the first digit.
- only protection against access to dangerous parts is indicated and the first digit is then replaced by an X.

Letter	Protection of individuals
A	back-to-hand access
B	finger access
C	tool access
D	wire access

Additional letter: material protection

Additional letters are used to provide additional special material information. They can be placed after the second characteristic digit or after the additional letter.

Letter	Material Protection
H	high voltage equipment
M	tested against the harmful effects of entering water by running equipment
S	tested against the harmful effects of entering water with equipment not in motion
W	suitable for use in specified atmospheric conditions

9.3.1 Equivalence of IP encoding with NEMA encoding

NEMA encoding (National Electrical Manufacturers Association) considers additional factors to IP encoding, such as corrosion protection.

NEMA encoding can be compared with IP encoding as shown in the table below. Please refer to the official manufacturer documentation of the sensor used for more details.

NEMA	IP	NEMA	IP
3	10	5	52
2	11	6	67
3R	14	6P	67
1	54	12	52
3S	54	12K	52
4	56	13	54
4X	56		

9.3.2   Classification for medical devices

AP Category Equipment

AP stands for Anesthetic Proof. Apparatus or part of apparatus conforming to the requirements specified for construction, marking and documentation in order to be protected from ignition of a mixture of flammable anesthetic and air.

Category APG apparatus

Appliance (Anesthetic Proof for use in 'Zone G') or part of the apparatus meeting the requirements specified for construction, marking and documentation so as to avoid flame sources in a mixture of flammable anesthetic and oxygen or nitrous oxide.

9.3.3   Level of equipment protection (EPL)

Level of protection granted to an electrical equipment based on its probability of becoming a ignition source according to EN 60079-0. See also ATEX explosive atmospheres paragraph.

The classification of equipment that can be used in explosive atmospheres differs according to whether it can be used in explosive atmospheres because of gas or dust and according to the level of protection offered:

- **Gb EPL**: equipment for use in explosive atmospheres for the presence of gas, with a 'high' level of protection, which is not a ignition source during normal operation or during expected malfunctions.
- **Gc EPL**: equipment for use in explosive atmospheres for the presence of gas, with an 'increased' level of protection, which is not a ignition source during normal operation and which has some additional protection measures to ensure that it remains an idle ignition source in the event of regularly expected events (e. g. lamp failure).
- **EPL Db**: explosive powdered combustible atmospheric equipment with a 'high' level of protection which does not constitute a ignition source in normal operation or when subject to expected failures.
- **Dc EPL**: explosive dust atmospheres equipment with an 'increased' level of protection, which does not constitute a ignition source during normal operation and which may have additional protections to ensure that it remains an inactive ignition source In the event of regularly expected events (e. g. lamp failure).

9.4   LASER EMITTERS

LASER (Light Amplification by Stimulated Emission of Radiation) is an electronic device, can be a diode, that converts energy into a thin, concentrated light beam. It can reach considerable distances.

The light produced by the LASER is an artificial light with special properties:
- Monochrome light, a single wavelength
- Directional
- High energy density
- Consistent emission (in phase)

Because of these characteristics the LASER can be used in many fields where another source of light would not be usable, for example in metal working or for surgical procedures.

The LASER beam can be obtained using the following means:
- Liquid (pigments)
- Gases (He-Ne helium neon, Ar argon, CO2)
- Solid (YAG Yttrium Aluminum Garnet, ruby, glass)
- Semiconductor (Gallium arsenide GaAs)

Unlike other types of radiation, LASER light has no cumulative effect on humans, although it is detrimental due to high energy, in particular to the skin and especially the eyes.

LASERs are categorized into increasing risk classes, depending on their power and hazardousness, according to the European standard IEC EN 60825-1 or American FDA 21 CFR Part 1040.10.

9.4.1 FDA Classification

CLASS	Indications
I	Not dangerous
II a	Radiation between 400 [nm] and 710 [nm]. When observed for less than 16 minutes, there is a chronic risk of damage beyond this time
II	Radiation between 400 [nm] and 710 [nm]. Radiation level considered chronic risk; protection is often given by spontaneous reflex of the eyelids
III a	Serious or chronic risk to vision when observed directly with an optical instrument
III b	Serious risk to skin and eyes in case of direct radiation
IV	Serious risk to skin and eyes in case of direct and diffuse radiation

A protective shield is required for all classes, as well as the identification plate. For classes IIIb and IV, a remote block command and a key command are also required.

9.4.2 IEC Classification

CLASS	Indications
1	Not dangerous, safe even in case of prolonged direct observation or optical instruments
1M	Safe even for long term vision, it can be dangerous when using lenses Radiation between 300 [nm] and 400 [nm]
2	Visible radiation between 400 [nm] and 700 [nm] Safe for temporary exposure, dangerous when fixing the light beam
2M	Visible radiation. Safe only for short exposure to the naked eye. It can be dangerous if you use lenses
3R	Potentially dangerous. Dangerous in case of intentional eye exposure
3B	Dangerous for direct observation, even accidental. The observation of diffuse reflections is normally safe
4	Dangerous to eyes and skin. The observation of reflections can be dangerous. Potential fire hazard

A protective shield is required for all classes, as well as the identification plate. For classes 3B and 4, a remote block command and a key command are also required. Class 4 requires manual rearmament.

The following is the hazard symbol to be applied near a LASER light sensor:

Common LASER light sensors used on machines are often classified in classes 1 or 2 according to the IEC EN 60825-1 standard and can therefore be used without additional precautions. These LASER sensors shall also display the warning symbol about the use of LASER light. The symbol is located on the sensor body or on the electrical connection cable.

Titolo | Sensors & Transducers - A practical guide 2023 edition
Autore | Emilio Carnevale
ISBN | 9791221485660

Youcanprint
Via Marco Biagi 6, 73100 Lecce
www.youcanprint.it
info@youcanprint.it
Made by Human